English Wetlands

A timely and important study of the liminal and untamed power of those delicate areas known as wetlands and how writers and artists have interpreted what they mean to us in an increasingly endangered world.
—Julia Blackburn Author of *Time Song: Searching for Doggerland*. Shortlisted for the 2019 Wainwright Prize and 2009 winner of the J.R. Ackerley award for *The Three of Us*

Rich and serpentine as the creeks it leaps over, English wetlands is sure to be regarded as a classic landscape work. Big scope and keen detail – It'll live in my rucksack for some time.
—Maxim Peter Griffin, contributor to *Caught by the River*, landscape artist, illustrator and filmmaker, author of *Field Notes*

Mary Gearey • Andrew Church
Neil Ravenscroft

English Wetlands

Spaces of nature, culture, imagination

Mary Gearey
School of Environment and
Technology
University of Brighton
Brighton, UK

Andrew Church
School of Environment and
Technology
University of Brighton
Brighton, UK

Neil Ravenscroft
Real Estate and Land Management
Royal Agricultural University
Cirencester, UK

ISBN 978-3-030-41305-7   ISBN 978-3-030-41306-4 (eBook)
https://doi.org/10.1007/978-3-030-41306-4

© The Editor(s) (if applicable) and The Author(s), under exclusive licence to Springer Nature Switzerland AG 2020
This work is subject to copyright. All rights are solely and exclusively licensed by the Publisher, whether the whole or part of the material is concerned, specifically the rights of translation, reprinting, reuse of illustrations, recitation, broadcasting, reproduction on microfilms or in any other physical way, and transmission or information storage and retrieval, electronic adaptation, computer software, or by similar or dissimilar methodology now known or hereafter developed.
The use of general descriptive names, registered names, trademarks, service marks, etc. in this publication does not imply, even in the absence of a specific statement, that such names are exempt from the relevant protective laws and regulations and therefore free for general use.
The publisher, the authors and the editors are safe to assume that the advice and information in this book are believed to be true and accurate at the date of publication. Neither the publisher nor the authors or the editors give a warranty, expressed or implied, with respect to the material contained herein or for any errors or omissions that may have been made. The publisher remains neutral with regard to jurisdictional claims in published maps and institutional affiliations.

Cover pattern © John Rawsterne/patternhead.com

This Palgrave Pivot imprint is published by the registered company Springer Nature Switzerland AG.
The registered company address is: Gewerbestrasse 11, 6330 Cham, Switzerland

Preface

The origins of this book are drawn from a research project entitled *WetlandLIFE: Taking the Bite Out of Wetlands*, which ran from July 2016 to July 2020, funded by the Valuing Nature Programme, supported by a number of UK Research Councils. The research team are from a consortium of UK universities alongside public and third-sector organisations and independent creative practitioners. The overall ambition of the project has been to improve wetland management by delivering national ecological guidance for managing insect populations, particularly mosquitoes, as part of healthy wetland environments, and to encourage the recreational use of wetlands to support the health and wellbeing of local populations. To do this, 12 English wetlands were selected for an ecological survey of mosquito species on each site. We then selected three of these case study sites, in Bedfordshire, Somerset and the Humber Valley, to research human use, experience, value and perception of local wetlands. This book draws on all elements of the project and other influences taken from our combined research portfolios concerned with water and the environment.

Thinking widely about the ways in which humans have shaped landscapes across time, and conversely the ways in which landscapes have sculpted human lives and cultures, this book seeks to celebrate the beauty, and complexities, of English wetlands from a contemporary purview. Attention is given to the ways in which these waterscapes have been much maligned, particularly in historical cultural representations, and how these waterscapes are increasingly understood as essential components for enabling transitions towards sustainable futures. Whilst the empirical fieldwork data which appears in all six chapters of the book is taken from the WetlandLIFE project work, the book also uses a range of materials drawn from other perspectives, including palaeoenvironmental archaeology; landscape architecture and environmental planning; human geography; ecosystem services; eco-criticism; literary, cultural and critical theory; environmental history; and natural resources management. As a result the book will appeal to a diverse audience. For those readers who feel an affinity with English wetlands and are keen to discover more about these spaces, the case study sites afford detail and nuance which are particular to these places and local communities, and also emblematic of wider changes and trends across these waterscapes at a generic level.

In order to provide clarity of focus, the book only reflects upon English wetlands, though readers are signposted to the work of other writers, both practitioners and theorists, throughout the book, for further reading and to enable considerations of other ways of 'knowing' wetlands that are outside the remits of this work. We hope that by the end of the book, our readers will feel sufficiently engaged and curious to visit the case study sites that we have so grown to admire and to feel encouraged to explore other wetlands, armed with novel insights which will hopefully enable them to view these landscapes in an entirely new way. So don your welly boots and get out into the great outdoors; you won't regret it.

Brighton, UK   Mary Gearey
                                                Andrew Church
Cirencester, UK   Neil Ravenscroft

Acknowledgements

The authors would like to thank our *WetlandLIFE: Taking the Bite Out of Wetlands* project partners. Our inimitable project leader Tim Acott receives a special mention for bringing together a wonderful group of researchers and for generating an enabling, inclusive and mutually supportive research environment. His colleagues from the University of Greenwich, Adriana Ford and Fakhar Raza, have been inspiring team-mates, and Mary and Andrew in particular thank Adriana for her enthusiasm and drive within the social sciences research they have developed together. We would also like to thank our other project colleagues: Frances Hawkes, Bob Cheke and Gabrielle Gibson from the Natural Resources Institute; Anil Graves, Joe Morris and Sharanya Basu Roy from Cranfield University; Peter Coates from the University of Bristol; David Edwards from Forest Research; Jolyon Medlock and Alex Vaux from Public Health England; and our creatives Victoria Leslie, Kerry Morrison and Helmut Lemke. We, of course, particularly thank our research participants for their part in developing the WetlandLIFE research project, and, now, this book.

Mary and Andrew would also like to thank their colleagues at the University of Brighton and especially those within the School of Environment and Technology.

Thanks from Mary also to Ian Mell, Angela Piccini, Bella Dicks, Cara Adair, Jane Moore, Mick Garrity, the Wotherspoons, Mannerings, Saunders, the Salts, Griffithses, Lewes pals and all the Barcombe team. I would like to thank all my family for their help, advice and encouragement. Benjamin Gearey's expertise has been greatly appreciated. My love and thanks to my two greatest constructive advisors, and children, Niamh Gearey and Arthur Gearey. My wonderful husband Adam Gearey has been a constant source of delight, and distraction, throughout the writing process. To Adam, Arthur and Niamh, my unbounded love.

Neil and Andrew would like to thank their family and friends for the support and to the many researchers and students who have worked with them over the years on water-related projects; Dr Johanne Orchard-Webb and Dr Becky Taylor have been valued colleagues over the years. In particular, Neil and Andrew are forever grateful to Mary for taking the lead on this ambitious project and drafting the majority of the text. That it has been completed at all is without doubt a tribute to her tenacity, while the breadth and quality of the arguments made are equally a tribute to her deep understanding of, and passion for, English wetlands.

Contents

1 Wetlands and Humans Across Time: An Overview   1

2 Wetlands in Depth: The Waterscapes of Bedfordshire, North Lincolnshire and Somerset   31

3 Wetlands as Ludic Spaces: Play, Recreation, Rejuvenation, In/Exclusion   67

4 Wetlands as Literary Spaces: Off Kilter, Off Grid, Off the Wall   91

5 Wetlands as Remembrance Spaces: Contemplation, Ceremony and Commemoration   119

6 Human-Nature Connectivity: Wetlands Within Sustainable Futures   145

Index   171

List of Figures

Fig. 1.1	Theoretical historical extent of English wetlands (indicative). (Credit: Wetland Vision Technical Document)	13
Fig. 1.2	Current extent of English wetlands (indicative). (Credit: Wetland Vision Technical Document)	14
Fig. 2.1	WetlandLIFE case study site locations. Red—Alkborough Flats; blue—Bedfordshire; green—Somerset Levels. (Credit: Mary Gearey)	33
Fig. 2.2	Map to show Priory Country Park and Millennium Country Park in Bedfordshire. (Credit: The Forest of Marston Vale Trust (adapted by Mary Gearey))	36
Fig. 2.3	Map detail of PCP layout. (Credit: Bedfordshire Borough Council)	38
Fig. 2.4	Map of Humber Estuary including Alkborough. (Credit: www.tide-project.eu) https://www.tide-toolbox.eu/abouttidetoolbox/abouttideproject/	47
Fig. 2.5	Shapwick Heath and Westhay Moor in relation to Glastonbury	58
Fig. 3.1	Guide to Shapwick Heath. (Credit: field-studies-council.org)	75
Fig. 3.2	Alkborough Flats Heritage trail: visitnorthlincolnshire.com	77
Fig. 6.1	SuDS. (Credit: Slow the Flow Calderdale)	157

List of Images

Image 1.1	Shapwick Heath, Somerset Levels, May 2018. (Credit: Adriana Ford)	4
Image 1.2	Mesolithic footprint from the Severn Estuary. (Credit: Professor Martin Bell)	6
Image 1.3	The night is a starry dome. (Credit: Tom Hammick, all rights reserved Bridgeman Images)	23
Image 2.1	WetlandLIFE team visiting Shapwick Heath in August 2017 with Mark Blake, reserve manager at SWT. (Credit: Tim Acott)	34
Image 2.2	Priory Country Park, commemorative plaque. (Credit: Mary Gearey)	36
Image 2.3	Priory Country Park, central lake zone, May 2018. (Credit: Mary Gearey)	37
Image 2.4	Plaque to mark Neolithic henge complex, Bedford's Riverside Tesco store. (Credit: Phil Shirley)	41
Image 2.5	Fenlake Meadows' information board, PCP, May 2018. (Credit: Mary Gearey)	42
Image 2.6	A PCP guide to 'the lake', May 2018. (Credit: Mary Gearey)	43
Image 2.7	Parakeets in an American Oak, Hyde Park, London, October 2019. (Credit: Ralph Hancock)	44
Image 2.8	The 'Timberland Trail', navigating the circumference of the Forest of Marston Vale, May 2018. (Credit: Mary Gearey)	45
Image 2.9	New lower embankment, Alkborough Flats, March 2018. (Credit: Mary Gearey)	48
Image 2.10	The cooling towers of the Drax power station, as viewed across from Alkborough Flats. (Credit: Mary Gearey)	49

Image 2.11	Information map at Alkborough Flats, March 2018. (Credit: Mary Gearey)	50
Image 2.12	Look-out post at Alkborough Flats. (Credit: Geography.co.uk)	53
Image 2.13	Werner 162. (Credit: Maxim Peter Griffin)	54
Image 2.14	Julian's Bower, Alkborough Flats' turf maze, February 2018. (Credit: Mary Gearey)	55
Image 2.15	Shapwick Heath's rich diversity. (Credit: Adriana Ford)	57
Image 3.1	The bog of eternal stench: still from the film *Labyrinth*	72
Image 3.2	Labyrinth, Priory Country Park, Bedford, January 2019. (Credit: Phil Shirley)	73
Image 3.3	Sweet Track reconstruction, Shapwick Heath, Somerset. (Credit: Avalon Marshes Centre)	76
Image 4.1	A classic front cover jacket for crime novel *The Hound of the Baskervilles* by Sir Arthur Conan Doyle	93
Image 4.2	The fearful Tiddy Mun. (Credit: Benjamin Gearey)	101
Image 4.3	Rusalkas. (Credit: theweirdandtheodd.com)	103
Image 4.4	The Beck Stone; part of The Stanza Stones Walk, West Yorkshire and Greater Manchester, UK. (Credit: Mick Melvin)	104
Image 4.5	Bungay's infamous Black Shuck town spire, Bungay, Suffolk, UK. (Credit: The Suffolk Coast DMO)	110
Image 4.6	Warrior Wolf Women of the Wasteland. (Credit: Carlton Mellic III)	112
Image 5.1	Reconstructed Ballachulish figure from the Pallasboy project, carved by Mark Griffiths. (Credit: Benjamin Gearey)	122
Image 5.2	Doggerland. (Credit: Maxim Peter Griffin)	124
Image 5.3	Alkborough parish church's stone labyrinth. (Credit: Mary Gearey)	126
Image 5.4	River lampreys. (Credit: https://commons.wikimedia.org/wiki/File:Lamprey_mouth.jpg)	128
Image 5.5	Memorial bench, Alkborough Flats, March 2018. (Credit: Mary Gearey)	131
Image 5.6	Family-orientated bat walks. Priory Country Park, May 2018. (Credit: Mary Gearey)	133
Image 6.1	Mai Po Wetlands, Hong Kong SAR, overlooking Shenzhen, China. (Credit: WWF Mai Po Wetlands)	146
Image 6.2	Mudskipper. (Credit: sfzoo.org)	147
Image 6.3	Border control within the Mai Po Wetlands, separating Hong Kong SAR from China. (Credit: Mary Gearey)	148
Image 6.4	Skull damaged by malarial anaemia. (Credit: Jess Beck)	153
Image 6.5	Feeding female mosquito. (Credit: Jolyon Medlock)	154

CHAPTER 1

Wetlands and Humans Across Time: An Overview

Abstract This chapter outlines the importance of wetlands as complex ecosystems which are fundamental to the health of the planet. We explore how human development is inextricably linked with these diverse landscapes. Much consideration is given to the way in which scientific and policy discourses around wetlands make use of generic evaluation frameworks such as the Ramsar Convention on Wetlands and, increasingly, ecosystems services perspectives. We critique how these shape and inform policy making in support of wetland protection, conservation and restoration. The chapter goes on to reflect on the use of wetlands by humans and more-than- human species across time, to broaden out our understanding of human relationships within these particular ecosystems. This opening chapter therefore sets the scene somewhat for more in-depth discussions developed throughout the course of the book.

Keywords Wetlands • Human development • Scientific and policy discourses • Evaluation frameworks • Ramsar • Ecosystem services

INTRODUCTION

This book invites the reader to consider the importance of wetlands to humans over deep time and the contemporary connectivity of human development alongside wetlands. Consideration is given to understanding that wetlands as places have specific socio-environmental impacts that vary spatially whilst recognising that temporality is also crucial when we reflect upon wetlands as spaces of nature, of culture and as landscapes shaped by the human imagination over time. We place special emphasis on interrogating the contributions wetlands make to wider physical, geographical and biotic processes, such as the global hydrologic cycle, and their importance as land and waterscapes which support biodiversity across varied terrains. Wetlands influence more than just their local environments. Their different functionalities extend beyond sovereign borders, as, amongst other attributes, they purify air, water and soil, regulate temperature and support migratory wildlife. As we will see as we progress through the book, wetlands, large and small, 'natural' and constructed, shape our everyday lives and have inter and intragenerational impacts on human health and wellbeing.

As a means to frame wetlands, throughout the whole book we explore tensions which exist between how different domains of knowledge, such as landscape planning, water resources management, ecology, legal frameworks and anthropology, amongst many others, challenge our understandings of what wetlands 'are' and how these perspectives directly influence how we use and value these land and waterscapes over different historical epochs, often to their detriment. Current paradigmatic definitions of wetlands, developed over recent decades, have collectively framed them as ecosystems which are singularly important for sustainability transitions. This encompasses the broad deployment of the term 'ecosystem services', shaping our appreciation of the importance of wetlands for human wellbeing and for the mitigation of climate change impacts. As we move through the book, we will be cognizant that human engagements with these spaces are always in flux, with dominant, hegemonic perspectives of what wetlands 'are', and consequently how they should be utilised, contested by a range of counterviews, actions and behaviours which, over time, shift our cultural practices. Wetlands may appear as tranquil, quiet spaces; as we will see, however, they are often the sites of radical, delinquent and subversive intent and so enable the incremental transformation of societies in many different ways.

Our focus in this book is on English wetlands, and in particular on three specific sites; in the West of England (Somerset), the East (North Lincolnshire) and the East Midlands (Bedfordshire). Our primary research, which we discuss more fully in Chap. 2, was undertaken in these locations, and an analysis of our data allows us to highlight the importance, and the relevance, of these landscapes for an understanding of the ways in which wetlands are hybrid spaces where nature, culture and the human imagination intersect. Informed by our case studies, the book aims to consider the importance of wetlands as fundamental to the health of the planet; this chapter sets out the key bodies of scholarship which contextualise our work so that we can achieve this ambitious aim. Existing discussions of wetland definitions and governance are an important context for our thinking. This chapter summarises how wetlands are defined and considers the governance regimes that shape wetlands, such as the Ramsar Convention, and its prescriptive taxonomies, so that we can explore what it means pragmatically in terms of protecting and rehabilitating wetlands. But the intellectual context to this book is much broader than governance, and this chapter also introduces the key scholarship that has influenced our thinking on wetlands, much of which critiques the ways in which nature is co-opted for human benefits alone and how the relations between humans and nature need to be radically rethought. Key political ecology arguments are explored through the work of Erik Swyngedouw and Donna Haraway, amongst others, to enable us to refine our deliberations concerning human-wetland relationships. Inherent to these analyses is the opportunity to really question our relationships with other living things on the planet—the other animals, the plants and all manner of biotic entities which together can be termed the 'more-than-human' (Haraway 2016). From this viewpoint, humans are just one element in a complex series of ever-changing interactions that we can call life on Earth. By using this method of moving between theoretical considerations and real-world case study examples, we wish to help you examine the wider dynamics at play in the shaping, use and value of our contemporary English wetlands, and thus to consider the wider global significance of wetlands. The book as a whole deals with a wide range of issues—land use management and modernity, local economies and livelihoods within nation-building endeavours, climate change mitigation and adaptation strategies within rural and urban contexts, individual wellbeing and community identities linked to place-specific sites, social and environmental justice issues, the role of culture in our connectivity with wetland landscapes. By considering scholarship, epistemologies and ontologies, this opening chapter therefore outlines key debates over the meanings, definitions, significance, value and governance of wetlands. This sets the scene for the later chapters which contain more in-depth discussions, often based on our three case studies, about time, space and wetlands (Image 1.1).

Image 1.1 Shapwick Heath, Somerset Levels, May 2018. (Credit: Adriana Ford)

Immersion into Wetlands

Humans and wetlands have been interconnected across time—deep time. Though deep time is a contentious term (Irvine 2014), with different uses across different disciplines, we can say that humanity's dependence on wetlands over millennia for all aspects of survival is incontestable (Schmidt 2017). As the land bridge between (what is now) the British Isles and continental Europe began to fill with river channels and marshland, our human ancestors left in memoriam their footprints and artefacts which enabled their survival—bone and antler spears (Leary 2015), flint axe blades (Van de Noort 2011) and evidence of Mesolithic structures (Momber 2011; Momber and Peeters 2017). Just as humans have always moved across the planet's surface, so have water bodies, land forms and other geomorphological processes. These are a reminder that rising and falling sea levels, glaciation and climate change are part of Earth's long history. It is essential though that climate cycles are understood as distinct from human-made, or anthropogenic, climate change. Climate change science clearly evidences that human-induced global warming is at a pace and scale far beyond natural climate variations (IPCC 2018). Wetland degradation across the globe over the last two hundred years is closely linked with the rapid rise of industrialisation and the marketisation of nature for profit. We need to step back and reflect on human reliance upon wetlands over millennia and consider that the places in which humans have chosen to live across the Earth are intimately connected to the migration of wetlands around the planet in response to geological time-scale adaptations:

> *Out from Cromer in an easy sea, Pilgrim Lockwood*
> *cast his nets and fetched up a harpoon.*
> *Twelve thousand years had blunted not one barb.*
> *An antler sharpened to a spike, a bony bread knife*
> *from a time of glassy uplands and no bread:*
> Greetings from Doggerland, *it said.*

Excerpt from *Doggerland*, Jo Bell.

Across the world, in different ways, humans over time have depended on wetlands for their lives and livelihoods—fish, fowl, plants and fungi for eating; reeds, wood and clay for shelter and tools; peat for fuel; pelts and skins for clothing; water for thirst and farming; medicines, art materials, rites and rituals all depend on the existence of these waterscapes. Wetlands are spaces for seclusion, safety, ceremony, storage and delinquency; all of which will be explored in more detail in following chapters. Using case study data and critical scholarship, we hope to stimulate curiosity concerning these unique habitats and their 'agency' in shaping an Earth including humans.

Often described as the 'kidneys of the landscape' (Mitsch and Gosselink 2015: 3) for their filtration and cleansing attributes, wetland metaphors connect with the human body in other ways too. Utilising the Netherlands' descriptor, to reflect the country's sea-level terrain, we might also think of wetlands as nether regions—low lying almost underworld spaces, crouched between sky and liquid earth, where the soft terrain and incumbent water meet human weight. In these regions, person and landscape become connected at ankle, knee, groin; sinking down. It's hard to remain upright on many parts of a wetland; when walking in wetland terrain, a human develops a rolling gait, like a sailor with sea legs—stumbling over tussocky turf and staggering across plains of water resting on the soil, the huge skies dancing off the reflection, creating reflected cloudscapes meeting the ground and sky. Traversing wetlands takes agility, a lightness of form, a close reading of the landscape (Image 1.2).

Humans depend on wetlands for far more than just local provisioning needs in terms of water, food, shelter and aspects of cultural practice and performance. The ecosystem services perspective, that has been promoted by the United Nations and many national governments over the last

Image 1.2 Mesolithic footprint from the Severn Estuary. (Credit: Professor Martin Bell)

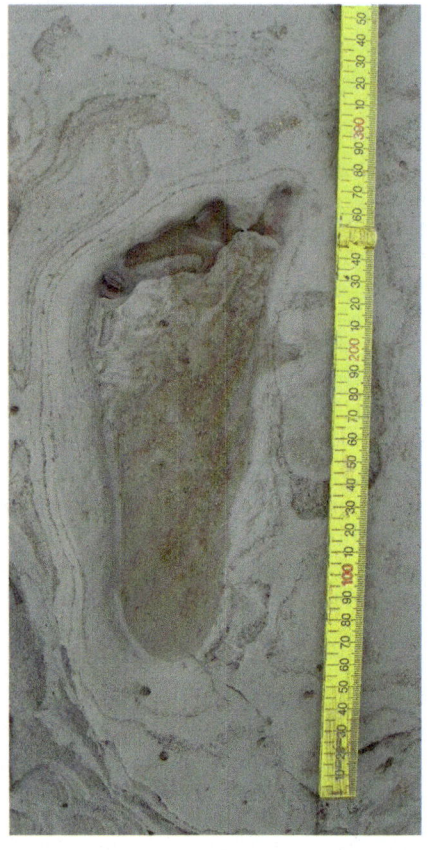

three decades, is a way of assessing and analysing the relationship between environments and human needs (Potschin et al. 2016). Ecosystem services approaches encourage us to understand the macro effects of wetlands on the planet, working across governance regimes and national boundaries and working with broader swathes of time. Different types of wetlands provide different ecosystem services. Some, such as peatlands, sequester carbon and riverine wetlands act as flood storage and retention spaces, capturing vast amounts of water from rivers in spate and from storm water. Wetlands help regulate air temperature, air quality and water recycling through the evapotranspiration processes of plants. They are an intrinsic facilitator of nutrient cycles and of the storage and release of both green carbon and blue carbon, sulphur, phosphates and nitrogen. They enable so much more. They provide breeding and feeding ground for pollinators; they protect and home migrating animals; they remove pollutants from water and soil (Mitsch et al. 2015).

The Slippery Definitions of Wetlands

Before a broader consideration of the meaning of wetlands to humans, it seems prudent to start with the ecological categorisation of wetlands and the types and extent of wetlands to be found currently in England. Broadly, wetlands are ecosystems within which there is partial inundation of water on a seasonal, stochastic or episodic basis (Mitsch and Gosselink 1993). These spaces can be pluvial, fluvial or tidally fed, by rain, river and seawater on the planet's surface, or saturated through underground channels—streams, springs and aquifers obscured to the eye. These waters can be freshwater, brackish or saline. Wetland systems include peatlands, such as fens and mires, coastal salt marshes, riverine floodplains and estuaries, lowland raised bog, blanket bog, purple moor grass and rush pastures. A sub-selection of open water habitats includes ponds, lakes and tarns as well as swales and retention basins (as drawn from the *Wetland Vision* technical document, 2019).

These are only partial descriptors, as wetland classifications in their specificity and complexity can cover whole books. Ecologists ascribe particular hydrological and physio-chemical attributes which can apply to most wetland spaces. As Holland et al. detail (1990: 172):

> Wetland ecosystems have one or three attributes: (1) they support, at least periodically, hydrophytes (2) the substrate is classified predominantly as an undrained hydric soil, and (3) the substrate is usually saturated with water or covered by shallow water at some time during the growing season each year.

As this book attempts to engage a wide readership, across disciplines, we use a classification of wetlands reduced to its most simplistic terms: that wetlands can be characterised as having water-loving plants and other biota; that they have slow-draining, water-abundant soil; and that surface water is present for specific and/or identifiable periods. Wetlands can be naturally occurring or artificially constructed by human agency and are, most abundantly, spaces which are shaped to some extent by land and water management techniques (Joyce 2012). 'Wetlands' are then slippery by definition. A dry woodland in summer months may still have the hydric soils and wetland plant species that make it a wetland—just that its appearance during more arid periods can prevent it from 'looking' like a wetland. The interface between wetlands and other ecosystems, for example, grasslands and peatlands, is gradual and to some degree subjective—and this fragility is described by ecologists as an ecotone (Gosz 1993). Ecotones are the areas of transition between differing ecosystem types; these are influenced by three factors in wetland identification: soil geochemistry, hydrology and the living entities (the plants, animals, fungi and bacteria) collectively described as the 'biota' that reside there, continually or peripatetically. These three interfaces can be both surface and subsurface, making the identification of some wetland systems hard for those who are not wetland ecologists (Tiner 2016). To some extent then there is a tension between precise scientific definitions of wetland environments and the subjective experience of wetlands (Greenland-Smith et al. 2016). As will be detailed across the following chapters, the social science research underpinning this book reveals a 'slippery' range of expectations from wetland users and visitors about what a wetland 'is' and, more specifically, what it should look like. Our perception and experience of landscape is enculturated: we have deep-seated expectations about what wetlands should look like and are scathing of spaces which differ from these wetlands of our imagination.

The 'we' throughout the book shifts and is as amorphous as a wetland itself. The 'we' refers to us as authors; at other times it represents the wetland users we have interviewed; it also represents the voices and attitudes of people responding to English wetlands on social media, through blog postings and other forms of written, visual or aural communication revealed through our research methodology. As authors 'we' are aware of the slippages of identity introduced in this work; by the end of the book, it is hoped that this playfulness and shape shifting has enabled our readers to feel both engaged with, and immersed within, English wetlands.

The Significance and Values of Wetlands

As outlined below, William Mitsch and James Gosselink's comprehensive work on global wetlands (2015) incorporates both changes to wetlandscapes over time and human agency as part of these changes. Their engaging and highly accessible text provides ecologists, geographers, and those interested in wetlands, scope to consider the importance and value of wetland management, both the conservation and restoration of these particular ecosystems, and of other ecosystems and global environments more generally. Mitsch and Gosselink's (1993, 2000, 2015) work is referred to throughout this book and is recommended to those wishing for detailed information concerning wetland ecology and management at a global scale, though with a North American focus.

The fifth edition of the Mitsch and Gosselink's (2015) *Wetlands* contains considerable new material compared to earlier editions, with four new chapters concerned with ecosystem services and wetlands, indicative of the current attention paid by researchers to wetland ecosystem services. The discussion of the intellectual history of ecosystem services emphasises that it is a twentieth-century approach, which by focusing on the value and values of nature for people seeks to highlight the failure of society to recognise its current and future dependency on nature (Potschin et al. 2016). Turner (2016) argues that the ecosystem services perspective has made a positive contribution to environmental management and that wetlands were one of the first ecosystems to be the focus of ecosystem services research. This highlights the diverse and multidimensional value of wetlands in terms of economic, ethical, religious, aesthetic and recreational uses.

A key contribution of the global scale ecosystem assessments is that they have identified clearly how wetlands are highly threatened spaces despite recent attempts at protection and restoration. The United Nations Millennium Ecosystem Assessment (MEA 2005a, b) aimed to take a global perspective on ecosystem services and the benefits that humans receive from nature, but it also focused on ecosystems of particular significance, including wetlands. The MEA (2005a, b) study that focused on wetlands and water identified major concerns at the global scale for wetland ecosystem services as the degradation and loss of wetlands were identified as being more rapid than was the case for other ecosystems. The study also projected that losses and degradation of wetlands in future would negatively affect human wellbeing, especially in lower income countries (MEA 2005a, b).

Just over a decade later, another global ecosystem assessment identified that the pressures on wetlands are still very pronounced. The United Nations Intergovernmental Science-Policy Platform on Biodiversity and Ecosystem Services (IPBES 2019) identified that natural wetlands are declining rapidly at 0.82–1.21% per year so that, in the areas studied, 31% of natural wetlands have been lost between 1970 and 2008 (Dixon and

Carrie 2016). This has consequences for key ecosystem services linked to wetlands as loss and degradation will release carbon from wetland ecosystems that collectively contain 12% of the global carbon pool (IPBES 2019). The Intergovernmental Panel on Climate Change (IPCC)'s 2018 report (IPCC 2018) has emphasised the critical importance of preserving wetlands as part of the global strategy to prevent warming by more than 1.5 degrees. Wetlands are then crucial to efforts to ameliorate the impacts of climate change. We need to go further. Recognising their importance has still not prevented wetland shrinkage and degradation at a global scale. A wide range of interacting factors have been identified as contributing to wetland loss and degradation, of which the most significant are population growth, infrastructure development, water extraction, land use change, pollution, eutrophication, overexploitation and invasive alien species. We explore these as we progress throughout the book.

Whilst these assessments provide a gloomy outlook for wetlands, they also emphasise that globally wetland restoration and conservation initiatives are increasing, especially when linked to bird protection and peatland management (IPBES 2019). Despite loss and degradation, wetlands still provide a wide range of beneficial ecosystem services including fish, fibre, water supply, water purification, coastal protection, climate regulation, flood protection, tourism and recreation (MEA 2005b). Also the many ecosystem assessments undertaken at the regional and national scales have identified how wetlands can be better managed and conserved in future. In particular, indigenous and local knowledge perspectives can play a crucial role in wetland restoration provided they are incorporated into the management process and not excluded as they have been in the past (Russi et al. 2013).

In addition, ecosystem assessments are concerned to identify policy options and have highlighted how integrated management approaches based on organisational and international collaboration are at the core of successful restoration. For example, in the last 30 years, the MedWet partnership between the European Commission, 26 governments, non-governmental organisations and the Ramsar Secretariat has resulted in 360 wetland sites in the Mediterranean Basin being designated under the Ramsar Convention as locations of international importance (Martin-Lopez et al. 2016). The importance of integrated collaboration and the incorporation of local knowledge for successful wetland restoration has also been highlighted in the UK by more populist writers drawing on an ecosystem services perspective. Tony Juniper, the former director of

Friends of the Earth UK, in the book *What Nature Does for Britain*, outlines how the management and flood protection of one of our case studies—the Somerset level wetlands—have, since devastating floods in 2014, been enhanced by partnerships between local residents, farmers, voluntary bodies, conservation groups, water companies and the governmental environmental management bodies.

A global momentum has built up around ecosystem services in academia and international environmental organisations as witnessed by the establishment of IPBES in September 2012, which now has 132 governments that support its ongoing activities. IPBES is seeking to do for biodiversity what IPCC did for the climate in terms of encouraging governments, business and society to respond to the threats arising from the ongoing degradation of nature. The work of IPBES is mainly done by academics contributing to the various international assessments which seek to synthesise existing peer-reviewed knowledge. One of the criticisms of the ecosystem services approach, however, is that it has yet to be adopted widely by local and even national environmental policy makers and actors (Noe et al. 2017; Waylen et al. 2015). Indeed, in our three case studies, ecosystem services were very rarely mentioned by the participants in our research. This is despite consistent attempts, including a national government White Paper, by the UK government since 2011 to establish the ecosystem services framework as a new approach to environmental policy at national and local levels (Waylen and Young 2014). A range of other criticisms and concerns regarding ecosystem services have been discussed in academic writing (Potschin et al. 2016). Criticisms have highlighted, in particular, the anthropocentric focus on the benefits that humans gain from nature, a concentration on economic value, rather than other forms of value and conceptual weaknesses. These include ignoring the 'dis-benefits' of ecosystem services which in the case of wetlands might include being breeding grounds for harmful pests such as mosquitos. The IPBES programme of work has sought to address these criticisms especially through significant conceptual and empirical work on different forms of value (Christie et al. 2019), the incorporation of indigenous local knowledge holders in the assessment process and adopting the more inclusive terminology of nature's benefits for people rather than ecosystem services (Díaz et al. 2018).

A study of English wetlands can't ignore these global considerations of the significance and value of wetlands based on an ecosystem services perspective as they are increasingly influential. Parts of this book seek to understand in detail some of the ecosystem services linked to wetlands, especially those with a cultural dimension. The remit of our book, however, aims to take a much more specific purview of wetlands which is broader than a traditional ecosystem services perspective. We do this by including analyses through the chapters which bring to the attention of the reader the ways in which contemporary politics influences the ways in which wetlands are embedded within English culture and the ways in which the ideation of wetlands shapes our use and values of these natural spaces right now. We frame our work spatially through a consideration of English wetlands, moving our lens of focus at the micro level through the use of case study data and panning out again to consider broader theoretical concerns around human-nature connectivity in a time of climate change. The final chapter of the book contemplates the contribution wetlands will make to a 'more than human' contribution to considerations of sustainable futures adaptations.

Wetlands, Sustainability and Governance

A connecting strand across all wetland scholarship is that these precious ecosystems are in decline and require management, conservation and protection. As can be seen in the two maps in Figs. 1.1 and 1.2, from the English cross-organisational campaigning group Wetland Vision, within the following chapters we'll be taking a closer look at drivers behind wetland degradation within England and examining current initiatives to restore, rehabilitate and expand these landscapes. Figures 1.1 and 1.2, from the Wetland Vision website, show how many wetland environments have been lost within England over the last two hundred years. Alongside the loss of habitat are the concomitant losses of animal and plant species, as well as spaces for human use and enjoyment.

The recognition of the need to control the impact of human interventions on land and waterscapes started in the UK from the late nineteenth century, with early attempts to control non-agricultural development often orchestrated by the Campaign to Protect Rural England (in its former guises). We must be attentive to the spatiality and temporality of anthropogenic intervention in English wetlands over millennia. Whilst our prehistory understandings of human agency are still developing, we can find clear evidence of Roman civil engineering in wetlands, particularly around the Severn Estuary, while the Fenlands are the clearest example of large-scale drainage schemes as initiated by the Dutch civil engineer Cornelius

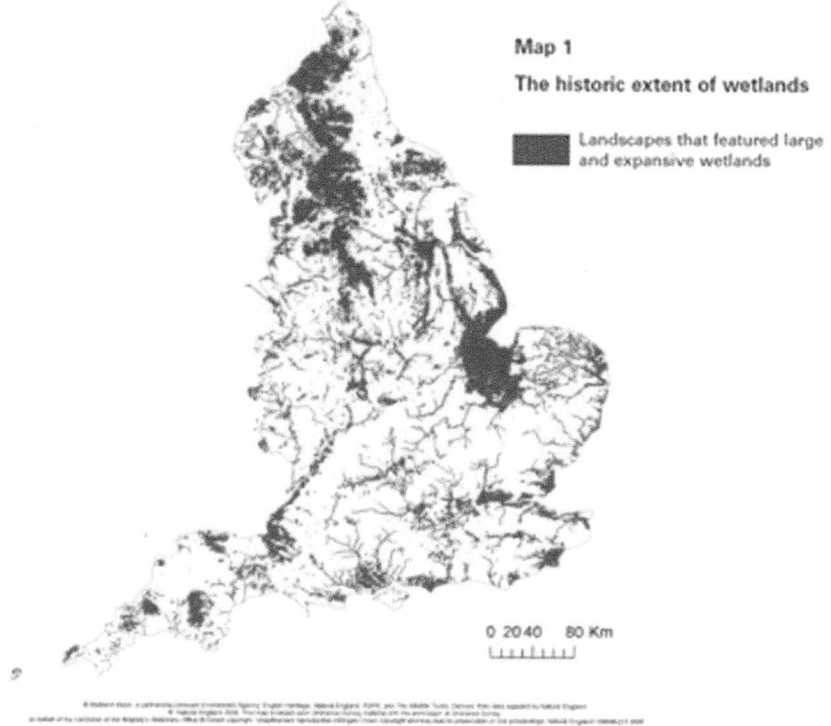

Fig. 1.1 Theoretical historical extent of English wetlands (indicative). (Credit: Wetland Vision Technical Document)

Vermuyden in the seventeenth century—which arguably contributed to the English Civil War (Hill 1970). Notable milestones included, for example, the Town Planning Act 1925, the Restriction of Ribbon Development Act 1935 and the Green Belt (London and Home Counties) Act 1938, all of which sought to control urban development and limit encroachment into the countryside. Much of this legislation was amalgamated in the Town and Country Planning Act 1947 which set out a process for mediating between public and private interests but did not provide strong protection for many wetlands and other fragile landscapes. At its core, the Act nationalised all development rights in land, meaning that landowners and occupiers could no longer decide unilaterally about the future use and development of their land and waterscapes. Rather, they had to seek permission to develop from

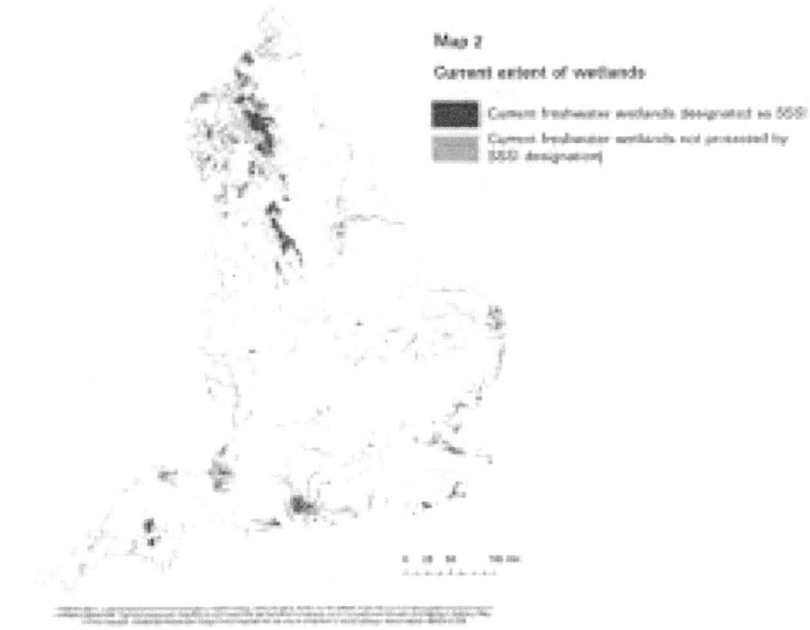

Fig. 1.2 Current extent of English wetlands (indicative). (Credit: Wetland Vision Technical Document)

newly established local planning authorities. This therefore provided a process for what Ravenscroft (1993: 116) has described as 'mediating between public and private interests' in any cases where proposals had been made for intensifying or changing the current use of land. The Act thus implied that, in at least some cases, socially and environmentally inefficient allocation of land uses would occur if it were not for the moderating influences of the planning system:

> The existence of town and country planning seems to be predicated on the assumption that there is some pattern of land use which is socially desirable, but which is different from the pattern which ... the market ... would produce. (Reade 1987: 3)

In its original guise, the mechanism for constraining development was based on an adversarial arrangement in which the local planning authority published a plan of what it wanted to achieve in terms of land use, and developers applied for permission to do what they wanted, whether or not this corresponded with the plan. The development plan thus became an indicative document that was understood as a material consideration when planning committees determined whether or not to grant permission to any specific planning application. In addition to this system, the Town and Country Planning Act 1947 and the National Parks and Access to the Countryside Act 1949 established a number of planning designations that sought to prevent most forms of development from taking place. These included National Parks, Areas of Outstanding Natural Beauty and Green Belts. The last of these, the Green Belts, have been particularly effective in preventing urban sprawl. However, none of these designations have had any element of coercion to improve or protect fragile, vulnerable or important land and waterscapes.

The National Parks and Access to the Countryside Act 1949 started the legislative process of protecting vulnerable spaces such as wetlands as it made provision for establishing National Nature Reserves (NNRs), which can be created by the statutory nature conservation agencies and managed by them or an approved body. Local authorities may also establish Local Nature Reserves (LNRs), provided the relevant statutory nature conservation agency approves. National Nature Reserves are used to protect some of the most important habitats, including wetlands, species and geology and to provide 'outdoor laboratories' for research. Most NNRs offer opportunities to schools, specialist interest groups and the public to experience wildlife at first hand and to learn more about nature conservation. There are currently 224 NNRs in England, covering approximately 0.7% of the country's land surface. Over half of these include some wetlands. According to Natural England, the largest NNR is The Wash covering almost 8800 hectares, while Dorset's Horn Park Quarry is the smallest at 0.32 hectares. Natural England manages about two-thirds of England's NNRs, with the remainder managed by the National Trust, Forestry Commission, Royal Society for the Protection of Birds (RSPB), Wildlife Trusts and local authorities.

Since the rash of post-war planning and land use legislation, there have been many attempts to amend and extend controls, including the Wildlife and Countryside Act 1981 and the Countryside and Rights of Way Act 2000, as well as considerable reform of the town and country planning system (2012). New regulations have also been incorporated in law as a result of European directives such as the Water Framework Directive (2000). Amongst the most significant of these for wetlands has been the designation of Ramsar sites across Europe, including the UK. As early as the 1960s, in recognition of the global decline in wetlands across the planet, leading scientists and environmental campaigners gathered together during a series of international meetings to agree how to uniformly protect wetland environments, across states, with widely varying resources to draw upon. As stated on the Ramsar website's opening page:

> The convention on wetlands, called the Ramsar Convention, is an intergovernmental treaty that provides the framework for national action and international co-operation for the conservation and wise use of wetlands and their resources.

The original convention was agreed in the city of Ramsar, Iran, in 1971. Redrafted several times since its original formulation, the Convention now states that its main goal is:

> the conservation and wise use of all wetlands through local and national actions and international cooperation, as a contribution towards achieving sustainable development throughout the world.

It's important to look at this statement in more detail. Global in ambit, the Convention places emphasis on a form of *noblesse oblige* custodianship—where emphasis is placed on retaining current wetlands through a suite of human responses that support 'conservation and wise use'. There is no guiding terminology for what this 'wise use' could entail: it is a highly subjective declaration. Whilst the Land Use, Land- Use Change and Forestry (LULUCF) section of the Kyoto Protocol suggests that 'wise use' of wetlands can count towards greenhouse gas mitigations, there is a presumption that renewable energy will steer humans away from encroaching on wetlands for fuel needs (Roulet 2000). This approach does not suggest how this will work in wetlands in peripheral economies nor addresses the environmental hazards of renewable energy technologies and components themselves (Haque et al. 2014; Bakhiyi et al. 2018). Further, the statement links wetlands directly with achieving 'sustainable development'. By linking wetland custodianship with an economic goal, sustainable development, the inference is clear: wetlands are productive zones, to be used and managed by humans for their financial benefit.

The Ramsar Convention clearly articulates an anthropocentric hierarchy in which humans are the main beneficiaries from these landscapes.

Unsurprisingly, the Convention has generated a host of critical responses. These include a lack of recognition, and so management guidance, regarding the dynamic range of wetlandscapes over deep time, what Gell et al. (2016: 875) call the 'nature and trajectory of change' of wetlands over thousands of years. Dixon and Carrie (2016) argue that the question of 'sustainable for who' should be addressed, as often not all wetland stakeholders are included in management discussions—particularly women or those from lower status local communities—as part of wider considerations about climate change, environmental degradation and the continued avarice that underpins economic growth, whether depicted as 'sustainable' or not.

Erik Swyngedouw's work (2007, 2010) asks us to question what he terms the oxymoronic qualities of the term 'sustainable development'. Developing this interrogation of the different modes of water management on the planet used for economic gain, Boelens et al. (2016) suggest that all water environments including wetlands should be recognised as hydrosocial territories: they are all, to some extent or other, managed for the benefit of humans and as a result are prone to manipulative political decision making. Asara et al. (2015) build upon this to suggest that only through rejecting sustainable development and moving towards shrinking economies, 'degrowth', can we truly protect livelihoods and planetary environments through understanding that harming nature harms humans too. We can see then that to fully understand and immerse ourselves in English wetlands, we must be receptive to approaching them with a curious, and critical, eye, especially in relation to how they are governed. Indeed the UK's own National Ecosystem Assessment found that the lack of a well-developed inventory of wetlands meant that policy was not based on a full understanding of the ecosystem services provided by wetlands in the UK, especially in relation to climate change (UK National Ecosystem Assessment 2011).

While the implementation of the Ramsar Convention involves an expectation of managed improvement of sites, the critique of the Convention can, to some degree, be levelled at all forms of government intervention in land use allocation decisions—that intervention is designed to support wider political objectives such as economic growth (including approaches which can be termed sustainable development). However, this is to view such interventions from a rather narrow perspective, when market interventions can achieve a range of objectives, for example, related to aesthetics, functionalism and the management of externalities. In many cases the justification for designating and protecting wetland environments has been on aesthetic and externality grounds: that wetlands improve the appearance of areas around cities, especially when these include abandoned industrial workings, as in the Bedford case study, and that the ecological benefits of wetland environments cannot be enjoyed by the public without some form of designation and governance. Equally, there are many functional reasons for protecting wetlands, related to their role in flood prevention, for example, as in the Somerset and Lincolnshire case studies.

Furthermore, many wetlands are owned privately but are made accessible through designation and access arrangements. These sites, increasingly referred to as POPS (privately owned public spaces), often incorporate wetlands, particularly in urban areas. These sites are often well used by local people who, according to Gearey et al. (2019: 409), 'effectively curate new civic subjectivities for themselves' through voluntary activity at the sites. In so doing, again according to Gearey et al. (ibid), they 'produc[e] for themselves new modes of "hydrocitizenship"'. These are hybrid forms of practice in which special interest groups and local communities 'accept' moral, if not actual legal, responsibility for the appropriate use of sites to which they are offered access but not ownership. As Gearey et al. (2019: 409) conclude:

> These emergent 'hydrocitizenships' thus encapsulate very specific geo-spatial subjectivities and performativities which lock in access to waterscapes with closely scripted conditionalities regarding activity and behaviour.

We must also consider then not just what wetlands 'are', but questions around access and use of these spaces. As developed in Chap. 6, social justice issues are a key concern as we think widely about sustainable futures and the place of wetlands within socio-environmental transitions.

Wetland Ontologies and Epistemologies (Extract from *Kinship*, Seamus Heaney 1974).

Quagmire, swampland, morass:
The slime kingdoms,
Domains of the cold-blooded,
Of mud pads and dirtied eggs.

But bog
Meaning soft,

The fall of windless rain,
Pupil of amber.

Ruminant ground,
Digestion of mollusc
And seed-pod,
Deep pollen-bin.

Earth-pantry, bone vault,
Sun-bank, embalmer
Of votive goods
And sabred fugitives.

Insatiable bride.
Sword-swallower,
Casket, midden,
Floe of history.

Ground that will strip
Its dark side,
Nesting ground,
Outback of my mind

As Heaney suggests, wetlands are difficult landscapes to love. As already alluded to, they are not the easiest of terrains to traverse—often boggy, mud slick, uneven, insect abundant, watery in all its elemental forms. Aside from buggy wet woodlands, treeless wetlands expose visitors to rain, wind, hail, snow and sun. Without prior knowledge of the terrain—of the moss, fungi, plants, insects and other animals, geomorphology and hydrology regimes of each site—they are spaces that are difficult to decode, hard to appreciate. Instead these less handsome spaces appear to be a morass of turf and tarn, while urban wetlands appear to be particularly raggedy spaces with demarcated areas for lake, reeds and willows. To the untrained eye, they are often boring—the stalking ground of camouflaged birders and anglers. Even the interpretation boards yield few clues about why we should care about these spaces, often listing plants and animals we might find during our wetland foray rather than connecting these spaces with wider environmental concerns or with human wellbeing. Wetlands aren't prone to self-promotion. You can't 'bag' a wetland; there's no personal best achieved from circumnavigating a bog; even the big skied fens are reliant on camera filters and cloudscapes to make them Instagrammable; picnics are hazardous events on a wetland. Yet notwithstanding these constraints, wetlands do prompt strong emotions, of joy, intrigue, frustration, rapture, resignation, bewilderment.

> What I love is one foot in front of another. South-south west and down the contours. I go slipping between Black Ridge and White Horse Hill into a bowl of the moor where echoes can't get out
> listen,
> a
> lark
> spinning
> around
> one
> note
> splitting
> and
> mending
> it
> and I find you in the reeds, a trickle coming out of a bank, a foal of a river. (Alice Oswald, Dart 2010: 2)

These amazing ecosystems, so varied and complex, linking mosaics of ever-changing land and waterscape across vast areas will, as the more instrumental ecosystem services perspective suggests, be an essential component of our human adaptivity to climate change impacts. Their ability to filter air and water quality, to store and release floodwater in response to their surrounding environments and to support biodiversity in animal and plant life, amongst many other attributes, makes them one of the most important ecosystem types on the planet. Yet as the Intergovernmental Science-Policy Platform on Biodiversity and Ecosystem Services (IPBES) assessment indicates globally they are still being degraded and lost despite attempts at protection.

As this book begins to outline all the different, and wonderful, ways that wetlands enhance human life, it is worth reflecting on the relationship between knowing a wetland and *knowing* a wetland. We must first make careful and considered steps—both to consider wetlands as landscape types, and so being able to identify wetlands and particularly those in an English context, and to reflect on our own individual connectivity with wetlands, drawn from personal experience both intimate and derived from books, films, media and other forms of cultural influence. This difference can also be described as the connection between ways of knowing wetlands—the forms of knowledge building and sharing that we can call epistemologies of wetlands—and the ways in which wetlands are incorporated into our life courses' habits and behaviours, what we may wish to describe as ontologies of wetlands. This is crucial as we begin to explore what wetlands 'are'. As we've seen wetlands have a scientific definition, given global resonance by the Ramsar Convention. But wetlands are slippery beasts; their dynamic constitution means that they move between states, much like water changes throughout its journey within the hydrologic cycle. Further, we also wish to suggest that wetlands are not a fixed entity for the humans, and more-than-humans, that are involved with them. Our enquiry is not to develop a closed compendium of ways in which English wetlands are experienced, used and valued. Rather our aim is to prompt you to consider your own views, perspectives and uses of English wetlands, to explore where wetlands fit within the natural world, our contemporary landscapes and societies' values for the Earth. We hope to encourage you to consider how policies, management processes and civic endeavours which impact wetlands have been shaped by cultural influences and representations and, in turn, how wetlands shape our social and cultural milieu. Finally, a key ambition is to engage you to explore the place of wetlands in our own imaginations—to see them as figurative as well as literal, and littoral, spaces. This book is then an attempt to tentatively explore both the epistemology and ontology of English wetlands, using data drawn from social science research undertaken within three case study sites between 2016 and 2019.

Wetlands, Meanings and the Self

As authors of this work, each of us has spent three years thinking about, reading about, talking about and wading within wetlands. We have brought our own experiences and considerations into this work—cycling on undulating bog roads in the Irish Midlands on holiday, watching starling murmurations in the Somerset Levels, boating on the Norfolk Broads as children. These experiences were our starting point for our own personal journeys into our three case studies. All the study sites are (re)constructed wetlands, though to vastly different degrees. The Somerset Levels sites of Westhay Moor and Shapwick Heath are flooded former peat workings though embedded within a wider chain of extant wetlands—so their presentation is to some extent timeless. The Bedfordshire sites of Priory Country Park (PCP) and Millennium Country Park (MCP) are former gravel pits, clay pits and landfill spaces—and look more sculpted than the Somerset sites. The Humber Valley site of Alkborough Flats is part of the North Lincolnshire coastal wetland realignment project and, due to its position at the confluence of the Rivers Trent and Ouse, feels less managed than the other spaces, though it is, of course, anthropogenically shaped due to its flood management role, as is the PCP site in Bedford. The research on these sites highlights that 'naturalness' is a valued element of wetland waterscapes. Specialist users such as birders and wildlife photographers, in particular, expect these sites to look 'less managed'. Wetlands then are desired wildscapes—places for types of nature interaction not possible in other spaces. For many the lack of on-site facilities such as toilets, a café or a visitor's centre is an attribute that reduces visitor pressure, leaving these spaces as the preserve of the specialist.

We also bring with us our positionalities, our own sets of expectations about researching wetlands, which have changed both through exposure to the sites and through working with economists, artists, entomologists and environmental historians as part of the wider research team. As outlined in the *Foreward* to the book, the research team was brought together through the WetlandLIFE project. Learning to see through another's gaze has been a fundamental shift in the way we have approached our own work. As a means to untangle the complexity of thinking through humans and wetlands, we have included in this chapter, as well as Alice Oswald's *Dart*, a long poem which captures the voices of those who reside alongside and within the River Dart, the work of poet Seamus Heaney. Drawing upon just one of his 'bog poems' as a way to immerse ourselves in our wetland journey has enabled us to think through a cultural politics of wetland spaces. Heaney's work across these series of poems captures the essence of our research interest in wetlands. Wetlands are specific spaces; their attributes are place specific, time specific. Change the hour, the day, the activity, the companion and the experience of being in the wetland—with its attendant set of understandings and emotions—will also alter. These sets of unique experiences combine to form a feeling, a mood, a response which the imagination conjures up when thinking about a particular place. Heaney's bog poems deal with the contemporary—watching/remembering his father dig turf, the discovery of a young man's body during the Troubles (Whyte 1991)—but also link these wetlands to deep time, the ancestors who traversed these spaces, the placement of bog bodies as sacrifice, trophy, votive in the long ago of antiquity. Heaney's wetlands are his wetlands, yet they are also universal wetlands—'Earth-pantry, bone vault, Sun-bank, embalmer' speaks to the ways in which humans have utilised wetlands throughout their time on Earth. Writers and other artists are drawn to these spaces (see Tom

Hammick's work, below). These special ecosystems provided sustenance as spaces rich with animal and plant life, as places to forage materials for shelter and tool making, as spiritual sites where the threshold between worlds has long believed to exist, as punitive spaces for punishment, as delinquent spaces for illicit activities, as contemplative spaces for reflection and more recently as health and wellbeing spaces to undertake recreation and sport (Image 1.3).

Image 1.3 The night is a starry dome. (Credit: Tom Hammick, all rights reserved Bridgeman Images)

We recognise the largesse involved in only discussing English wetlands. English wetlands are related hydrogeomorphically with other wetlands across the UK; ecosystems do not adhere to political boundaries. Scale is everything for wetlands. As Mitsch and Gosselink (1993: 201) explain:

> Regional wetlands are integral parts of larger landscapes—drainage basins, estuaries. Their function and their values to people in these landscapes depend on both their extent and their location.

Some wetland functions and particular ecosystem elements exist only in dynamic relation to other landscape regimes such as forests, grasslands or rivers. Further to this the interconnectedness of wetlands with other global ecosystems is evidenced through the seasonal migratory patterns of birds. The mysterious traveller to our wetlands in Bedfordshire and Lincolnshire, the bittern, migrates from Europe to North Africa and Asia, depending on the species, throughout the year to inhabit other wetland spaces. The springtime cuckoo (though not a wetland native has welcomed us on each site with its distinctive call) overwinters in the rainforests of the Democratic Republic of Congo. We see then that when we consider English wetlands, we need to be always cognisant of their interconnectedness with other wetlands, other landscapes. In many ways this repositions wetlands from being dormant, domestic waterscapes with a bleak charm, to what Timothy Morton has described in his work *Dark Ecology* as 'hyperobjects' (2016). Hyperobjects are those climate and landscape entities whose importance is too abstract, immense or overwhelming for us to easily grasp. We can think of wetlands as hyperobjects due to their massive distribution across time and space. If we were to run changes on the Earth's surface as a film, we would see both the impacts of climate on wetland systems and, running at a faster tempo, the impacts of human agency on wetlands. Wetlands' hyperobjectivity can be ascribed due to this deep-time, globalised interconnectivity, meaning that these landscapes require us to consider the political ecology aspects of human-nature relationships (Larsen 2016). Their 'objectification' comes through anthropocentric intervention which renders them an artefact—something shaped by human endeavour. We can only begin to consider hyperobjects in small comprehensible pieces. Their scale and intensity makes grasping their totality fleeting, awkward, overwhelming. For instance, rather than fully open ourselves to the true horrors of climate change and the complete usurpation of all ways of life that we know to be 'normal', our human reaction according to Morton is to approach it and rationalise it in smaller ways. We can consider cycling to work, or turning off the lights at home, or taking fewer international flights as examples. We attempt to assert our agency in ways that do not disrupt our everyday.

We can apply this same hyperobject principle to wetlands. Wetlands have an agency to support life, not just human life, on the planet in numerous ways—providing thermal stability, water purification, soil regeneration amongst others—which go beyond national borders. Wetlands' influence on air quality, the hydrological cycle, food provisioning and more are global in scale, with temporal modes which operate over the everyday and also longer time cycles. The loss of global wetlands outside of England impacts English wetlands, through declining animal species and consequent impacts on local food chains, even if these impacts are indirect and slow to metastasise. Edgeworth and Benjamin (2017) apply this analysis to the Chicago River, USA, as an anthropogenically influenced hyperobject entity; we do likewise through an analysis of English wetlands. Throughout the book, we ask you, the reader, to consider not just the locality, the specificity of the case study sites we work with, but also their relationship with the wider issues at play: growing urbanisation and the loss of natural space; climate change impacts and our adaptation, mitigation and transformational responses; the marketisation of nature as part of contemporary neoliberalism.

As part of thinking around wetlands and humans across time, we must also consider the enmeshed relationships between the two. Wetlands are assemblages; no two wetlands are the same. In any given season or period of time, each wetland will have a slightly different assemblage of animal and plant life, bacteria and fungi. Temperature and water flow rate will change—as will its physical look, its chemical components. Yet these ever-changing assemblages enable the 'co-functioning' (Anderson and McFarlane 2011) of all the individual agents and of the wetlands in its totality. As humans then we are linked in a relationship of kinship with nature, where we are not the only components to have agency—or full control. Our shared future lies with connecting with nature, adapting and transforming, as we too are assemblages that change and adapt in what Donna Haraway describes as 'tentacular' responsive relationships with our more-than-human brethren (Haraway 2016). As Anderson and MacFarlane reflect: 'Assemblages are constantly opening up to new lines of flight, new becomings' (2011: 126). This recognition usurps the latent governance presumption, as epitomised by the Ramsar Convention, that human agency dominates what 'happens' to wetlands and that their role is to provide ecosystem services. Climate change impacts are beginning to fully reveal just how far nature determines human life. We continue this refrain in Chapter 2 . Looking more closely at our three English case study sites, we delve more closely into both the unique characteristics of these waterscapes and how the forces which have shaped their current management regimes can be seen to be historically and politically charged.

REFERENCES

Anderson, B., & McFarlane, C. (2011). Assemblage and geography. *Area, 43*(2), 124–127.

Asara, V., et al. (2015). Socially sustainable degrowth as a social–ecological transformation: repoliticizing sustainability. *Sustainability Science, 10*(3), 375–384.

Bakhiyi, B., et al. (2018). Has the question of e-waste opened a Pandora's box? An overview of unpredictable issues and challenges. *Environment international, 110*, 173–192.

Bell, J. (2015) A climate change poem for today: Doggerland by Jo Bell. The Guardian, 21st May 2015. https://www.theguardian.com/environment/2015/may/21/a-climate-change-poem-for-today-doggerland-by-jo-bell As accessed 5th August 2019.

Boelens, R., et al. (2016). Hydrosocial territories: a political ecology perspective. *Water International, 41*(1), 1–14.

Christie, M., et al. (2019). Understanding the diversity of values of Nature's contributions to people: insights from the IPBES Assessment of Europe and Central Asia. *Sustainability Science, 14*, 1267–1282.

Díaz, S., et al. (2018). Assessing nature's contributions to people. *Science, 359*, 270–272.

Dixon, A., & Carrie, R. (2016). Creating local institutional arrangements for sustainable wetland socio-ecological systems: lessons from the 'Striking a Balance' project in Malawi. *International Journal of Sustainable Development & World Ecology, 23*(1), 40–52.

Edgeworth, M., & Benjamin, J. (2017). What is a river? The Chicago River as hyperobject. *Rivers of the Anthropocene*, 62–175.

European Union. (2000) Directive 2000/60/EC of the European Parliament and of the Council of 23.10.00 Establishing a Framework for Community Action in the Field of Water Policy–EU Water Framework Directive (OJ L 327, 22.12.2000).

Gearey, M., Ravenscroft, N., & Church, A. (2019). From the hydrosocial to the hydrocitizen: water, place and subjectivity within emergent urban wetlands. *Environment and Planning E: Nature and Space, 2*(2), 409–428.

Gell, P. A., Finlayson, C. M., & Davidson, N. C. (2016). Understanding change in the ecological character of Ramsar wetlands: perspectives from a deeper time–synthesis. *Marine and Freshwater Research, 67*(6), 869–879.
Gosz, J. R. (1993). Ecotone hierarchies. *Ecological applications, 3*(3), 369–376.
Greenland-Smith, S., Brazner, J., & Sherren, K. (2016). Farmer perceptions of wetlands and waterbodies: Using social metrics as an alternative to ecosystem service valuation. *Ecological Economics, 126*, 58–69.
Haraway, D. J. (2016). *Staying with the trouble: Making kin in the Chthulucene.* Durham. USA: Duke University Press.
Haque, N., et al. (2014). Rare earth elements: Overview of mining, mineralogy, uses, sustainability and environmental impact. *Resources, 3*(4), 614–635.
Heaney, S. (1974). Kinship. *James Joyce Quarterly, 11*(3), 227–237.
Hill, C. (1970). *God's Englishman: Oliver Cromwell and the English Revolution.* London: Weidenfeld and Nicolson.
Holland, M. M., Whigham, D. F., & Gopal, B. (1990). The characteristics of wetland ecotones. In H. Décamps & R. J. Naiman (Eds.), *The ecology and management of aquatic-terrestrial ecotones* (Vol. 4, pp. 171–198). Florida, USA: CRC Press.
IPBES. (2019). *Global assessment report on biodiversity and ecosystem services of the Intergovernmental Science-Policy Platform on Biodiversity and Ecosystem Services.* Bonn, Germany: IPBES Secretariat.
IPCC. (2018) Global Warming of 1.5 °C: An IPCC special report on the impacts of global warming of 1.5 °C above pre-industrial levels and related global greenhouse gas emission pathways, in the context of strengthening the global response to the threat of climate change, sustainable development, and efforts to eradicate poverty. Summary for Policymakers (IPCC SR1.5). Released October 6, 2018.
Irvine, R. (2014). Deep time: an anthropological problem. *Social Anthropology, 22*(2), 157–172.
Joyce, C., 2012. Preface: Wetland services and management. *Hydrobiologia, 692*(1), 1–3.
Larsen, S. C. (2016). Regions of care: a political ecology of reciprocal materialities. *Journal of Political Ecology, 23*(1), 159–166.
Leary, J. (2015). *The remembered land: surviving sea-level rise after the last Ice Age.* London: Bloomsbury Publishing.
Martin-Lopez, B., et al. (2016). Ecosystem services supplied by Mediterranean Basin ecosystems. In M. Potschin, R. Haines-Young, R. Fish, & R. K. Turner (Eds.), *Routledge Handbook of Ecosystem Services* (pp. 405–415). Abingdon: Routledge.
Mitsch, W. J., & Gosselink, J. G. (1993). *Wetlands* (2nd ed.). New York: John Wiley.
Mitsch, W. J., & Gosselink, J. G. (2015). *Wetlands* (5th ed.). New York: John Wiley.

Mitsch, W. J., & Gosselink, J. G. (2000). The value of wetlands: importance of scale and landscape setting. *Ecological economics, 35*(1), 25–33.

Mitsch, W. J., Bernal, B., & Hernandez, M. E. (2015). Ecosystem services of wetlands. *International Journal of Biodiversity Sciences, Ecosystems Services and Management., 11,* 1–83.

Momber, G. (2011). Submerged landscape excavations in the Solent, southern Britain: climate change and cultural development. In J. Benjamin et al. (Eds.), *Submerged prehistory* (pp. 85–98). Oxford: Oxbow Books.

Momber, G., & Peeters, H. (2017). Postglacial human dispersal and submerged landscapes in north-west Europe. In *Under the Sea: Archaeology and Palaeolandscapes of the Continental Shelf* (pp. 321–334). Switzerland: Springer.

Morton, T. (2016). *Dark ecology: For a logic of future coexistence.* Columbia: Columbia University Press.

Noe, R., et al. (2017). Mainstreaming ecosystem services in state-level conservation planning: progress and future needs. *Ecology and Society, 22*(4), 4. https://doi.org/10.5751/ES-09581-220404.

Oswald, A. (2010). *Dart.* London: Faber & Faber.

Potschin, M., et al. (2016). Ecosystem services in the twenty-first century. In M. Potschin, R. Haines-Young, R. Fish, & R. K. Turner (Eds.), *Routledge Handbook of Ecosystem Services* (pp. 1–11). Abingdon: Routledge.

RAMSAR. https://www.ramsar.org/ accessed 020919.

Ravenscroft, N. (1993). *Recreation planning and development.* Basingstoke: Macmillan.

Reade, E. (1987). *British town and country planning.* Milton Keynes: Open University Press.

Roulet, N. T. (2000). Peatlands, carbon storage, greenhouse gases, and the Kyoto Protocol: Prospects and significance for Canada. *Wetlands, 20*(4), 605–615.

Russi, D., et al. (2013). *The Economics of Ecosystems and Biodiversity for Water and Wetlands.* IEEP, London and Brussels: Ramsar Secretariat, Gland.

Schmidt, J. J. (2017). *Water: Abundance, scarcity, and security in the age of humanity.* New York: New York University Press.

Swyngedouw, E. (2007). Technonatural revolutions: the scalar politics of Franco's hydro-social dream for Spain, 1939–75. *Transactions of the Institute of British Geographers, 32,* 9–28.

Swyngedouw, E. (2010). Impossible sustainability and the post-political condition. In M. Cerreta & V. Monno (Eds.), *Making strategies in spatial planning: Knowledge and values* (Vol. 9, pp. 185–205). Dordrecht: Springer.

Tiner, R. W. (2016). *Wetland indicators: A guide to wetland formation, identification, delineation, classification, and mapping.* Florida: CRC press.

Turner, R. K. (2016). Economics and ecosystem services: a positive contribution to environmental management. In M. Potschin, R. Haines-Young, R. Fish, &

R. K. Turner (Eds.), *Routledge Handbook of Ecosystem Services* (pp. 115–118). Abingdon: Routledge.

United Nations. (2005a). *Millennium ecosystem assessment, ecosystems and human well-being: Synthesis.* Washington DC: Island Press.

United Nations. (2005b). *Millennium ecosystem assessment, ecosystems and human well-being: Wetlands and Water.* Washington DC: Island Press.

Van de Noort, R. (2011). *North Sea archaeologies. A maritime biography, 10,000 BC–AD 1500.* Oxford: Oxford University Press.

UK National Ecosystem Assessment. (2011). The UK national ecosystem assessment: synthesis of the key findings. UNEP-WCMC, Cambridge. Available from: http://uknea.unepwcmc.org/Resources/tabid/82/Default.aspx.

Waylen, K., Blackstock, K., & Holstead, K. (2015). How does legacy create sticking points for environmental management? Insights from challenges to implementation of the ecosystem approach. *Ecology and Society, 20*(2), 21. https://doi.org/10.5751/ES-07594-200221.

Waylen, K. A., & Young, J. (2014). Expectations and experiences of diverse forms of knowledge use: the case of the UK National Ecosystem Assessment. *Environment and Planning C: Government and Policy., 32*(2), 229–246.

Wetland Vision. http://www.wetlandvision.org.uk/dyndisplay.aspx?d=maps_present accessed 020919

Whyte, J. (1991). *Interpreting Northern Ireland.* London: Clarendon Press.

WEBSITES

https://www.visitscotland.com/see-do/active/walking/munro-bagging/ accessed on 020919

CHAPTER 2

Wetlands in Depth: The Waterscapes of Bedfordshire, North Lincolnshire and Somerset

Abstract This chapter begins to interrogate the themes outlined in Chap. 1 by making use of data drawn from the WetlandLIFE case study sites. An overview is given of the English research locations: Shapwick Heath and Westhay Moor in the Somerset Levels, Priory and Millennium Country Parks in Bedfordshire, and Alkborough Flats in North Lincolnshire. To understand these contemporary wetlands in relation to English landscapes, the chapter undertakes a socio-historical review of each of these ecosystems, exploring how these spaces have been anthropomorphically reshaped up until the present day. The themes include 'drain and reclaim' in support of modern state building, nature based solutions and green infrastructure, wetland grabs and green gentrification, and the importance of urban wetlands as restorative spaces for human health and wellbeing.

Keywords Shapwick Heath • Westhay Moor • Somerset Levels • Priory Country Park • Millennium Country Park • Alkborough Flats • Drain and reclaim • Nature based solutions • Green infrastructure • Wetland grabs • Green gentrification • Restorative spaces • Wellbeing

This chapter provides an introduction to three English wetland networks that typify some of the wetland regimes found across Northern Europe. As detailed in Chap. 1, wetland landscapes cover a range of differing topographies and hydrologies. As stated, our concern in this book is not to provide a systematic taxonomy of the classifications of wetlands found in England, but rather to consider how wetland systems have shaped our regional and national culture and how culture has shaped how we use and value wetlands. As a means to explore this throughout the book, this chapter will outline the three case study wetland sites that were researched as part of the WetlandLIFE project and which underpin the empirical fieldwork utilised in this book. These three sites were selected for their importance in representing the variety of wetland type, form and function that exist throughout the country; we use them to consider the past and present factors which have, and do, impact on their use and value to humans and the more-than-human (Haraway 2015). Categorising wetlands by taxonomic landscape system alone does not capture the co-dynamic that has always existed between these waterscapes and the human management and governance regimes and practices which have sculpted them as biophysical entities. We must also consider that the ways in which these waterscapes have been modified or used over time reflect wider discourses within different social, historical, political and economic epochs. Wetlands have always been 'used' by humans. Whether in provisioning terms for food and shelter or for cultural reasons such as ritual, play, or for health, humans have always utilised wetlands for their own benefits, using context-specific ontologies and epistemologies. Addressing these multiple, complex and often contrary forms of use is a central ambition of this book. Reflecting on the ways in which we have revered, engineered and renaturalised our English wetlands over antiquity to the present day enables us to appreciate how central they are to us as lived and worked landscapes (Ingold 1993; Massey 2004) and how deeply they have influenced us culturally as spaces of beauty, reflection and rejuvenation, as well as spaces of otherness and foreboding. Wetlands, like other terrains, exhibit the contrariness of our human relationships with nature. We understand our reliance upon environmental systems for our existence on the planet, yet still treat them like commodities which can be bargained, exchanged and offset. We use these case studies to think about the complex and contested co-relationships between humans, nature and cultural practices in wetland spaces (Fig. 2.1).

Priory Country Park and Millennium Country Park in Bedfordshire are constructed wetlands; they are human made. Constructed wetlands vary in size and location, from small artificial filtration beds in new building developments to reclaimed post-industrial spaces in city centres and urban hinterlands. Our Bedfordshire sites are examples of urban and peri-urban

Fig. 2.1 WetlandLIFE case study site locations. Red—Alkborough Flats; blue—Bedfordshire; green—Somerset Levels. (Credit: Mary Gearey)

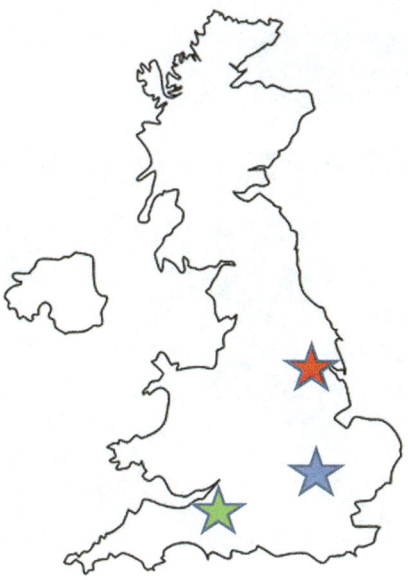

constructed wetlands that provide flood storage facilities and are also accessible, aesthetically 'less managed' natural spaces for local residents and visitors to enjoy. Alkborough Flats, our case study site on the Humber Estuary, is an example of an estuarine constructed wetland. This wetland is also an example of arable reversion, where the flood defences of riverside farmland have been deliberately breached to reinstate floodplain that would have originally characterised this section of the catchment. The wetlands here are part of a wider chain of riverine, estuarine and coastal wetlands collectively termed the Humberhead Levels. Finally, the third case study area comprises two connective wetlands within the Somerset Levels: a wide expanse of low-lying peatlands, comprising salt marshes, fens and raised bogs. The Somerset Levels are part of a wetland system that can be dated back 4500 years; we see that human agency has always shaped these spaces. From the Neolithic 'Sweet Track' found on Shapwick Heath to the current regime of carefully maintained water levels to support resident and migratory wildlife, managed through sluices and river gates also present in Westhay Moor, these wetlands are as artificial as the constructed wetlands of Bedfordshire. All three case studies alert us to the degrees of intervention, or management,

Image 2.1 WetlandLIFE team visiting Shapwick Heath in August 2017 with Mark Blake, reserve manager at SWT. (Credit: Tim Acott)

of these diverse wetland systems: nature, culture and imagination are connective aspects within English wetlands (Image 2.1).

Bedfordshire: Priory Country Park and Millennium Country Park

Bedfordshire is a gem of a county. Tucked tightly between its larger neighbours, Northamptonshire and Cambridgeshire, and only some 50 miles from London, Bedfordshire manages to offer a diverse array of habitats within its 123,000 hectares (Bedsfire 2019). A mainly agricultural county, sloping gently into Fenland at its east, rising to the Dunstable Downs and Chiltern Hills at its south, Bedfordshire opens out into the English heartland to its west. Its close proximity to London has been central to its economic development. Bedford's post-war wealth was built on mineral extraction, light manufacturing and engineering, partly stemming from its aeronautics heritage. Its relative geographic centrality and proximity to road and rail networks enabled its market gardens and distribution hubs to easily service the whole of the UK. During the Second World War, many children were evacuated to Bedfordshire and, in the post-war period, London families moved to Bedford, Leighton Buzzard, Dunstable and the newly developing Milton Keynes, bringing their love of West Ham United football club, the Hammers, along with them.

The county is currently experiencing quite rapid developmental change. This is shaped by two forces. The first is the cascade effect of high property prices in London. Trains from Euston to Bedford take under 40 minutes, meaning that Bedford and its surrounding villages are commutable and affordable. This has led to rapid new residential developments across the county. Secondly, Bedford is key to the development of the 'Oxford-Cambridge Arc' (MHCLG 2019). This planning and development initiative seeks to reconnect the Oxford to Cambridge train line, enabling quicker transit to both cities from London via Bedford. This is also prompting property acquisition and speculation within the county.

Just as Bedfordshire's economy and population distribution has been shaped by changing forces, so too have the two wetlands selected as our case study sites. Both wetlands are urban or urban fringe and are sited on reclaimed aggregate pits. Bedford Priory Country Park (PCP) is located within the centre of the city of Bedford, and Millennium Country Park (MCP, which is located within the emergent Forest of Marston Vale) is six miles south of Bedford in an area of rapidly developing new housing within and surrounding the villages of Stewartby and nearby Marston Moretaine (see Fig. 2.2). In a small county undergoing great pressures to develop on greenfield sites, conservationists and environmentalists have lobbied hard to both protect green spaces within the county and to ensure that there is access to 'wilder' green spaces for its growing, and increasingly urbanised, population. This makes our two Bedfordshire case study sites extremely important to their local populations.

Priory Country Park is the older of the two wetlands. Established as a nature reserve in 1977, after the end of industrial-scale gravel extraction on the site, it officially opened to the public in 1986 (see Image 2.2). It sits adjacent to the River Great Ouse and acts as an overflow floodplain and water meadow to prevent Bedford from flooding in times of high river levels. With 360 acres of floodplain, lakes, dry and wet meadows, riverside walks and small copses, the park is managed in such a way that it is able to reconcile the needs and expectations that people associate with a typical inner-city park with its status as a wildlife reserve. It accomplishes this through careful spatial planning, where parking, cemented paths, signage,

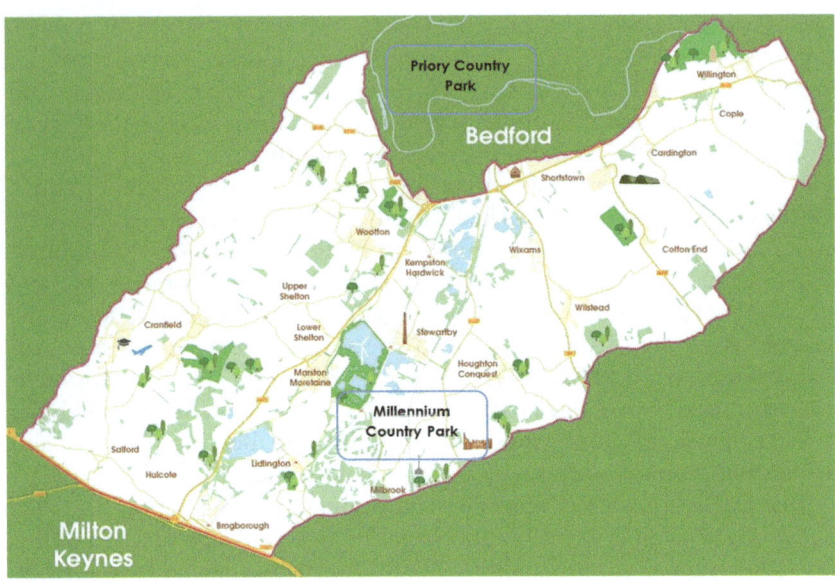

Fig. 2.2 Map to show Priory Country Park and Millennium Country Park in Bedfordshire. (Credit: The Forest of Marston Vale Trust (adapted by Mary Gearey))

Image 2.2 Priory Country Park, commemorative plaque. (Credit: Mary Gearey)

café, bird hides, play park and toilets are focused around the main Priory Lake and boating marina (see Image 2.3 and Fig. 2.3). These unconscious nudges direct users towards remaining within the central 'honeypot' area, only 100 metres from the main car parking area. Most self-guided activities occur within this zone which constitutes a relatively minor footprint within the Park. As we can see from Fig. 2.3, Priory Lake and marina only cover around 40% of the total area. The strategy from BBC has been to enable less managed spaces within the Rough, New Meadow, Fenlake Meadow, the Finger Lake area and the section which runs along to the Riverside Meadow (see Fig. 2.3).

These attributes have the effect, according to the park rangers, of keeping most of the visitor pressure to these spaces. Here the rangers undertake work to keep the grasses mowed, trees pruned or pollarded, and waste bins emptied. This deliberate strategy of greenspace management in a selected area serves to disincline most visitors from accessing other areas

Image 2.3 Priory Country Park, central lake zone, May 2018. (Credit: Mary Gearey)

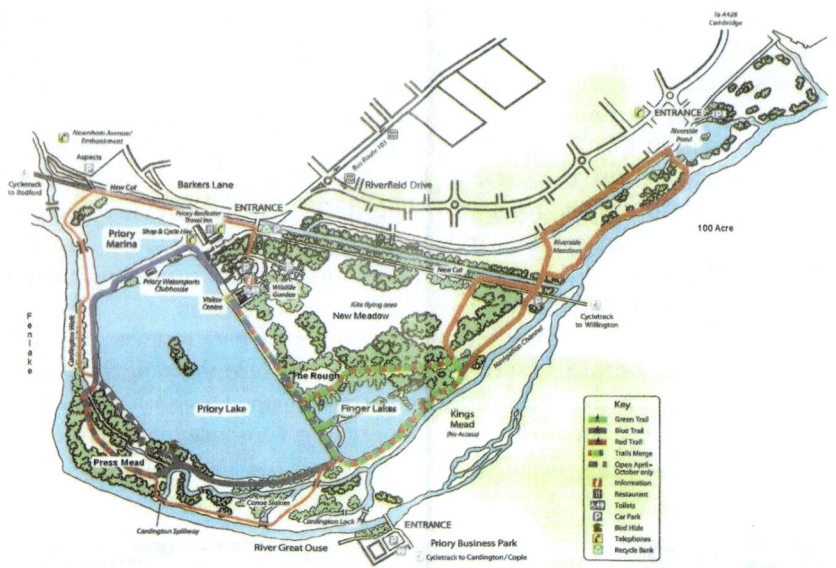

Fig. 2.3 Map detail of PCP layout. (Credit: Bedfordshire Borough Council)

of the site which the team wishes to leave unhampered in order to protect wildlife—animal, plant and aquatic. In these spaces meadows are left to self-seed, some areas are fenced to protect beehives or breeding sites, and footpaths are left rugged and rutted. Explored more below and in Chap. 3, this 'soft management' approach is not just important for leaving spaces unfettered, and to deter footfall, but is also a necessary strategy as council cuts have recently reduced the number of full-time salaried rangers at PCP from three to one. This mirrors a pattern across the UK (Heritage Lottery Fund 2014).

Priory Park is owned and managed by Bedford Borough Council, so decisions made regarding maintenance and use of the park are made by the elected local councillors. This governance regime contrasts with that found at Millennium Country Park, which is owned and run by a charity, with decisions made by its trustees. Like many English municipal green spaces and their attendant wetlands, Priory has been gravely impacted by austerity cuts, vastly reducing the amount of money available to spend on the site. As greenspace provision is discretionary and not statutory (Mell 2018), UK local authorities often cut the budgets for these activities first.

In England between 2008 and 2018, local councils have had to reduce their budgets by a national average of 23% (Gray and Barford 2018), and their spending on municipal green spaces as part of 'environmental and regulatory services' in the periods between 2010–11 and 2016–17 has been cut by £910 million (ibid: 558).

To replace the rangers who have been made redundant, the council have outsourced landscaping roles such as mowing, tree cutting and reed bed removal to third-party sub-contractors. Through outsourcing this landscaping work, the environmental management regimes are fixed, with a number of pre-agreed visits at designated intervals, slightly more during the growing months over spring and summer (Dempsey et al. 2016). The rangers we spoke to conveyed the inadequacies of this 'top-down' approach to greenspace management. They spoke of the importance of looking holistically at the site during each season and over a year, to determine sensitive cutting and planting regimes and to monitor changes in plant and site health. As climate change and seasonality also impact growth cycles and migrant populations of species, cutting at pre-appointed slots may erase breeding sites, kill fledglings or inhibit self-seeding plants, which may have a cascade effect on other sites within the wetlands and to other wetlandscapes within the county.

Austerity politics in the UK has had impacts on these wetlands in other ways as well. Income generation becomes increasingly important to cash-deprived councils, and finding sources of income is viewed as entrepreneurial and responsive to funding constraints. As a result many councils have hosted activities on these greenspace sites, either generating income from the events themselves (proms in the park, live music events) or linking with commercial enterprises (e.g. zip wire firms and visitor attractions such as butterfly houses) or the cross-promotion from charity events—through uptake in café spend or income generated through car parking fees.

Given the lack of security for future park funding and municipal green-blue spaces, reports have speculated that new management models for urban parks are needed in the future (Barker et al. 2017). One vision for Priory is to network it with other green wetland spaces along the River Great Ouse to create a 'Bedford River Valley Park', (http://www.bedfordrivervalleypark.org.uk) starting at Priory and extending seven miles out of the centre. Creating this strategic vision is a means to protect smaller local wetlands through collective unity.

The location of Priory Country Park (PCP) poses some interesting questions around functionality and accountability which are equally applicable to many urban wetlands across England. Alongside its identity as a municipal green-blue space, Priory has a fluvial flood risk management role which is a statutory duty to be managed by the Environment Agency, the UK government regulator for larger watercourses. PCP provides flood storage for times when the River Great Ouse is in spate. The River Great Ouse's water level is highly affected by rainfall events higher up the catchment. Whilst the floodplain does not extend across all of Priory Park, it is a significant element of its riverside characteristic. In addition, Bedford—like many towns and cities—is becoming more vulnerable to pluvial (rainfall) flooding, as a result of changes in the intensity of storm events and also by the increased hardscaping of cities, as more grey infrastructure prevents soil absorption of rainfall and less green infrastructure prevents water uptake via evapotranspiration. The green footprint of municipal spaces such as Priory Country Park is as important in flood prevention as is their blue storage components.

Heritage also plays an important part of the Priory Country Park site development alongside its hydrological functionality. Named after the Priory ruins that still form the site's northern boundary, the in situ lakes (Priory and the Finger lakes) were formed from old worked gravel beds. As a consequence this is a post-industrial wetland, with its landscaping part determined by the riverine system and shaped by its aggregate and mineral mining past. On the site itself, there is little mention of the importance of this geographic space in supporting the growth of Bedford. In many ways this information is erased. Only by walking with a ranger on the site in spring and seeing differential growth patterns in the grass is it possible to discern the footprints of the two cooling towers of the coal-fired Goldington power station at the eastern end of the park, which was demolished in 1987.

This is not the only lost, or disguised, heritage on the site. After the demolition of the power station came further redevelopment of the adjacent area, including the purchase of Bury Farm next to the site. Later developed into a Tesco superstore, it was discovered that a Neolithic henge had been built there, along with a processional causeway that most certainly would have been linked to the river (bedsarchives 2019). As this development predated compulsory archaeological works, which are now statutory components of Environmental Impact Assessments, this site is

2 WETLANDS IN DEPTH: THE WATERSCAPES OF BEDFORDSHIRE, NORTH... 41

Image 2.4 Plaque to mark Neolithic henge complex, Bedford's Riverside Tesco store. (Credit: Phil Shirley)

now lost. The entranceway to the Tesco superstore records the lost henge, but the park has no mention of this riverside space as an important Neolithic site (see Image 2.4).

Erasing humans from English wetland landscapes as a means to recast them as non-human, or rather, pre-human, spaces is an important facet of wetland lore in reclaimed industrial landscapes. Memories of the space as brownfield sites are overwritten through references to early, pre-industrial incarnations. Their authenticity, or veracity, as natural spaces is attested to through reference to their longevity across multiple human generations. In the case of Priory Country Park example signage attests to the historical importance of the meadows for overflow flood relief and for seasonal drainage (see Image 2.5).

These information boards focus on 'nature', but to the exclusion of humans (see also Image 2.6). Instead, humans do not inhabit the natural world as nature, but as observers of nature or as governance actors. There

Image 2.5 Fenlake Meadows' information board, PCP, May 2018. (Credit: Mary Gearey)

is an inherent surveillance command/control element to these artefacts, to remind the casual bypasser that this is a recreational zone to be enjoyed, thoughtfully, by humans, with some spaces kept isolated for wildlife.

These boards are targeted towards assumed predictable, knowable user groups. The language is in English and has not been updated or tagged to direct walkers to online resources which might support those with English as a second language or those with different physical, mental and emotional requirements. Wildlife is cast in a benign framing. Foxes, rats, rabbits, wild mink, magpies, pikes—all scavengers, foragers and cannibals of the animal world are erased from these urban wetlands. This is true too of domestic creatures that populate nearby open spaces, domestic cats and dogs, released snakes and lizards, escaped parakeets (which now have established populations in some areas of London's Hyde Park, see Image 2.7)—these animals are too urban and so do not fit within the paradigm of the urban wetlands. The message is that the urban wetlands are 'wild' and humans manage the degree of 'wildness' they should perform.

2 WETLANDS IN DEPTH: THE WATERSCAPES OF BEDFORDSHIRE, NORTH... 43

Image 2.6 A PCP guide to 'the lake', May 2018. (Credit: Mary Gearey)

This concept of relative wildness and rewilding also connects with Millennium Country Park. The Forest of Marston Vale, within which the Millennium Country Park is located, is an example of a Community Forest. Community Forests are a global phenomenon (Bishop 1991), seeking to increase woodland spaces, particularly in peri-urban areas. In the UK, the most common management and governance structure for Community Forests is a community partnership, usually between public, private and charitable landowners with support and guidance provided directly or indirectly by local government. Since the 1990s Community Forestry has been an initiative of what is now Forestry England, seeking to communicate the importance of forests for ecological, economic and psycho-social benefits. The Forest of Marston Vale is just such an intervention, with the target of increasing land cover by trees to 30% by 2050. Eight miles of land between Bedford and Marston Vale was earmarked as a site for a Community Forest, and the Forest of Marston Vale was initiated in the year 2000. Though the

Image 2.7 Parakeets in an American Oak, Hyde Park, London, October 2019. (Credit: Ralph Hancock)

designated area covers 61 square miles, tree cover in 2000 was only 3% of the area: two-thirds less than the national average for open ground (Forest of Marston Vale Trust 2008). The forest is situated in post-industrial brownfield sites that were formerly brickworks and areas of land that are now covered landfill. Through the planting of over five million trees in a 20-year window, it is hoped to improve the ecosystem functionality of these tracts of land, uniting them visually through congruent tree species, to make an aesthetic contribution to the landscape. Signage around the Vale encourages walkers to circumnavigate the space (see Image 2.8) and presumes both a lack of familiarity with the area and with walking. The intent is clear; this is a post-industrial space that has been reclaimed for recreation and health to encourage those who never access the countryside and who do not view these reclaimed spaces as such.

Negotiating the different needs and expectations of the different landowners has played an important part in how this intervention has evolved, with Millennium Country Park being the 'flagship' site of the 12 individual

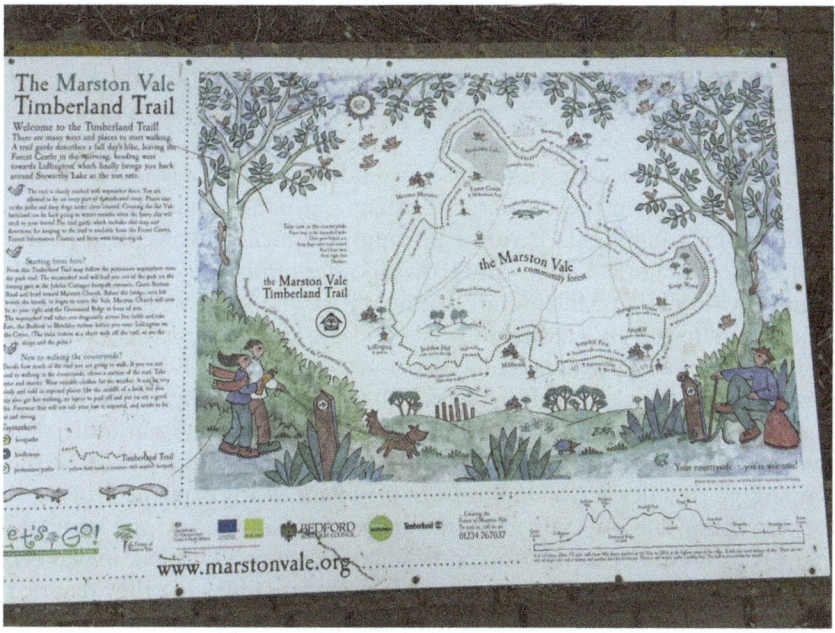

Image 2.8 The 'Timberland Trail', navigating the circumference of the Forest of Marston Vale, May 2018. (Credit: Mary Gearey)

sites which together make up the Forest of Marston Vale. The intent behind Millennium Country Park was to develop it as the focus space, to encapsulate the ethos of the Forest. The site is dominated by the Forest Centre, a modern visitors' centre which is multifunctional. Its café space is well used, and it also acts as a space for community events such as conferences, celebratory dinners, information days, fund-raising events and as an art and cultural space. Millennium Country Park is self-financing, and the Forest Centre activities, its attendant car park, family activity days and its 'Wildlife reserve' walk which charges a small fee all finance the rangers and landscaping work in the park and wider Forest of Marston Vale resourcing.

Community Forests can be viewed as a pragmatic route to make environmental change happen (Lawrence and Ambrose-Oji 2015), particularly in financially straightened times. Returning to our earlier theme of wetland management and resourcing, this model enables long-term planning and is not reliant on ongoing financing from central government or local councils. In many ways Millennium Country Park represents an emergent form of wetland development whereby the costs of managing these sites are no longer borne by the landowner but by the charitable trust and its management group. This model is seen more widely across the UK and is true for our other case study sites, through the creation of management boards, or community partnerships, often in the hands of Wildlife Trusts.

Tensions arise though with public misunderstandings of land ownership and management in these emergent urban and peri-urban wetlands. The signage which highlights Bedford Borough Council's involvement in the project (see Image 2.8) is in conflict with other information on site which reminds visitors that Millennium Country Park and the Forest of Marston Vale are managed by charitable trusts and are thus reliant on donations and other forms of private income generation. Given that Bedford's Priory Country Park has clear Bedford Borough Council branding and free parking, public expectations at Millennium Country Park are naturally that the space is a municipal space funded by council revenue. Until 2017 the car parking facilities at Millennium Country Park clearly explained the need for donations. Despite this few visitors paid for parking and the management team were reluctantly forced to impose charges. This led, inevitably, to a short-term dip in visitors as many felt affronted by what they deemed stealth taxes by the council. Patient and persistent reiterations by the Millennium Country Park management team and rangers have led to better public understanding of the financial operations of the site.

This links in many ways to public debates around access to the 'countryside'—who it is for and how it should be managed. Addressed more comprehensively in Chaps. 3 and 5, access to wetland spaces in England is concerned with many things: confidence in being in nature, what 'to do' in nature, having transport to these spaces, who you visit these spaces with, and perceptions of risks and safety in these less intensively managed spaces. There are a variety of needs and expectations around wetland spaces with a concatenation of 'Goldilock' responses. Some have too much signage, whilst others do not have enough; some are deemed as dominated by visitor centres with an 'exit through the gift shop' approach, others do not have enough facilities for visitors; for some the on-site parking promotes the 'wrong type' of users who visit the space to let their dogs off leads and have no interest in the wetland space itself; for others parking needs to be improved to allow access for all ages, incomes and mobilities. Wetlands, like other geographic spaces, reflect back to us the contemporary social values and representations that we prioritise as we move through the first quarter of the twenty-first century.

North Lincolnshire: Alkborough Flats

Sited some three miles up-river from the Humber Bridge, Alkborough Flats is the most contemporary of our three case study wetlands. It is a prime example of coastal realignment, where farmland has been allowed to flood through the breaching of natural, and constructed, river flood defences. The rationale behind this return to natural inundation at the confluence of the rivers Trent and Ouse, forming the Humber Estuary, has been to provide flood protection to the cities of Hull and Goole, further downstream. The Humber Basin (the hydrological area where the Ouse and Trent drain into a shared catchment known as the Humber Estuary) drains over 20% of the land mass in England (Humber Nature Partnership 2020) and see Fig. 2.4.

As the surrounding topography is low lying, only two metres above the high tide mark, the coastal ports, and associated urban areas, which are economically dependent on this water channel are prone to flooding in extreme weather events. In the case of Alkborough Flats, predicted sea level rise and increased fluvial and pluvial flooding events made the need to find nature based solutions (NBS) to support flood preparedness

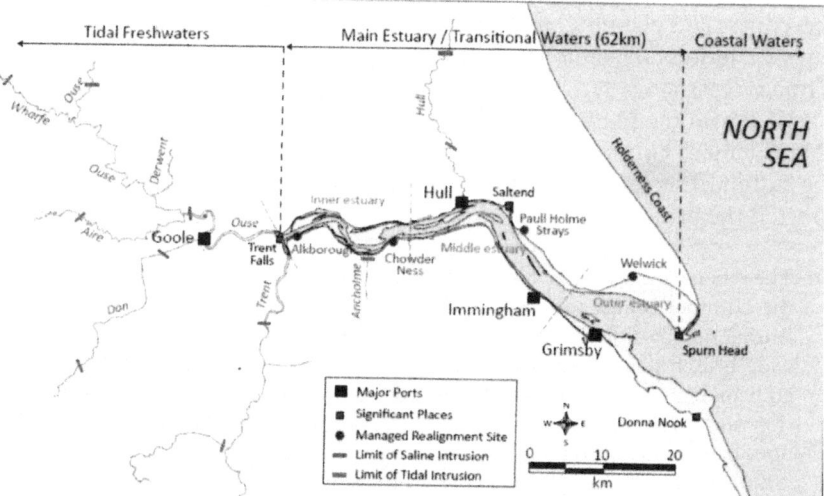

Fig. 2.4 Map of Humber Estuary including Alkborough. (Credit: www.tide-project.eu) https://www.tide-toolbox.eu/abouttidetoolbox/abouttideproject/

Image 2.9 New lower embankment, Alkborough Flats, March 2018. (Credit: Mary Gearey)

necessary. Alkborough Flats is then a classic example of multi-agency preparedness and planning, using blue-green infrastructure to develop strategies to address dynamic land and water responses to a changing climate (Image 2.9).

Works on the site began in 1999 under the purview of the Alkborough Management Group, a consortia of stakeholders from across private, public and third sector organisations. As the JBA Consulting Group who supported the project (2011: 4) state:

> The scheme increases the level of flood protection to an area stretching from the Humber Bridge to Goole up the tidal River Ouse and as far as Keadby Bridge on the tidal River Trent. The scheme features include a 20 metre wide breach in the existing flood bank, a 1,500 metre length of lowered embankment or spillway (see Image 2.9), new habitat areas, a pumping station and a new section of flood bank to protect assets. The breach in the floodbank adjacent to the Humber Estuary was made on the 6th September 2006 and the first tidal waters entered the site on the 8th September 2006.

Image 2.10 The cooling towers of the Drax power station, as viewed across from Alkborough Flats. (Credit: Mary Gearey)

The resulting wetlands are now intertidal habitats for migrating and local wildlife. The wetland regime is then a mix of fresh and saline water due to the influence of both riverine systems and a tidal estuary. Reed beds now cover the area of the breach, threatening to block the long eyeline vista that initially spanned across the wetlands and towards the banks of the Ouse and the Trent. Two car parks were created for access by visitors, and a concrete path links one end of the flats with the other. Two bird hides are on site, one with disabled access (Image 2.11).

Unlike the other case study sites, Alkborough Flats is a large industrial landscape. The Drax power station (Image 2.10) dominates the landscape, generating 7% of the UK's electricity and supporting nearly 3000 local jobs (https://www.drax.com/sustainability/people). Both the Ouse and the Trent carry huge cargo ships, both outward and inward bound. This site is not bucolic in its strictest sense. Its beauty takes a little time to recognise. Alongside the light playing across the river systems, and the reed beds

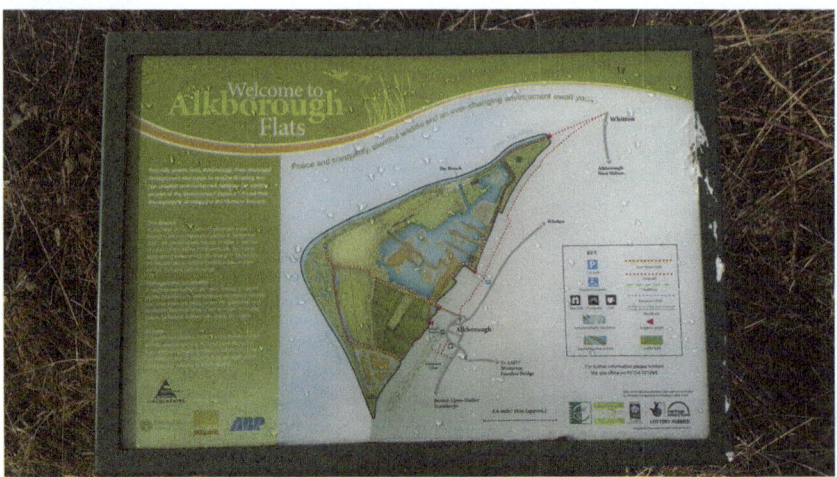

Image 2.11 Information map at Alkborough Flats, March 2018. (Credit: Mary Gearey)

swaying in the breeze, there are the funnels of the Drax cooling towers billowing steam and the muted horns of the river traffic. The A15 is a white noise of car and lorry drone as a bass note to the calls and shrills of the water birds. At 440 hectares in size, the nature reserve is hugely beneficial in supporting local wildlife and attracting seasonally migrating water birds.

It is worth reflecting on the scale and costs of the works within the context of the physical and human geography of the area. The Alkborough tidal defence scheme was completed in 2006, costing £10.2 million. This figure includes costs for compulsory land purchase as well as the hard landscaping. Part of the Shoreline Management Plan for the Humber Estuary, this wetland creation is just one element of an ongoing and proposed complex of natural flood management nodes along the Humber Basin.

Alkborough Flats' development represents a contemporary approach to natural flood management (NFM). NFM approaches have developed great momentum over the past 15 years (Holstead et al. 2017). In NFM intrusive and expensive infrastructure to prevent flooding is replaced with a changed mindset that recognises that flooding is inevitable and, indeed, natural and that solutions lie in working in harmony with natural systems. These include wetlands to store and release floodwater, wide riverine banks with undeveloped floodplains, mixed land use and tree cover to prevent soil erosion and enable water percolation within the soil structure. Here, the interventions are argued to be 'sustainable' as land use management is altered to recognise the intrinsic ability of natural systems to store and slowly release flood and storm water.

Yet within this purview, or NFM paradigm, rural spaces are unequal partners with the urban spaces they protect. Natural environments from this ecosystems perspective provide a service to urban centres. Although the benefits of NFM to natural spaces are plentiful, such as supporting local biodiversity as well as a range of other provisioning and regulating roles, we should be mindful of the agency behind these changes. It is the needs of the urban environments which determine the shape, scale and pace of the rural adaptations. There is a population bias within the NFM approach—with the urban still dominating over the rural. As Cook et al. suggest (2016), the paradigm of sustainable flood management (SFM), which is in many ways an interchangeable set of terms and conditionality with NFM, has a propensity to become co-opted into more entrenched, traditional, technical approaches, just couched under a different, more populist name. What we wish to be mindful of here is not that NFM or SFM are bad or unwanted, clearly not, but in their current application they do manifest modes of governance which infer ingrained traits regarding agency and decision making in land use management policies with an urban rather than rural bias. The needs and desires of rural populations are presumed less important, or at least less pressing, than those of their urban counterparts.

It is worth pursuing this a little more as we consider the historical, economic and demographic context of Alkborough. The area around the wetlands is rural, with the local village, Alkborough, typical of the smaller settlements which populate the large and relatively disconnected county of Lincolnshire. The village's name has had several incarnations and was listed in the Domesday book as Alchebarge, meaning 'the ridge like cliff above the mooring pool of the river' (Wittering 2017). The village sits above the riverine/estuarine stretch on a natural bluff, part of 50-mile escarpment called the Lincolnshire Limestone series or Jurassic Ridge. The area in which, Alkborough sits is known as 'Trent Falls'. As a result, the village itself has never been prone to flooding. Any interventions to reduce flood risk further down the estuary would have no obvious benefits for the village itself. In many ways the deliberate breaching of the 1950s built riverine flood defences could be designated an active example of 'managed retreat' now openly an Environment Agency policy, which has been effectively implemented, albeit quietly, over the last decade or more

(Harrabin 2019). This approach typifies a response to climate change issues along the East Coast of England, with many coastal villages reluctantly accepting that sea defences are no longer cost effective or preventative. Some low value areas will need to flood to protect, or enable resources to be directed to, more economically important spaces.

Questions do arise though, around how much government policy is influenced by a coordinated, vocal network of citizens. This is more challenging in disaggregated rural spaces. Alkborough and its parish has a population of around 460 people. The village is typical of this part of North Lincolnshire, with a local population that is mainly white, an average age of 50 plus years and little local employment. Housing stock is mainly privately owned. The largest town close by is Scunthorpe, eight miles away, described by former resident John Payne (2017) as an 'industrial island', a small but compact heavy manufacturing economy, rooted in the largely rural expanse of Lincolnshire. From the fieldwork undertaken with village residents, many of the incoming retirees of the village have relocated from Scunthorpe, Doncaster and Sheffield for their own rural idyll. Having talked to residents and local parish councillors, it was clear from the case study research that the development of Alkborough Flats was viewed positively. The creation of a footpath and bird hides still provides residents and visitors with an easily accessible space for walking, often with dogs, and for those with limited mobility, the well-maintained and flat footpaths were hugely beneficial. Yet many residents were disappointed that more had not been done to promote the wetlands. Some locals mentioned that, as part of the redevelopment scheme, they had understood that funding had been secured for a visitor centre. Others were disappointed that there was not more signage on the main roads leading to Alkborough to attract visitors. Poor public transport was also stated as a reason as to why other surrounding wetland spaces had higher visitor numbers. Some of the walking groups we talked to stated other reasons for a lack of take-up.

The 'Nev Cole Way' (https://www.walkingenglishman.com/ldp/nev-coleway.html) was developed by the Wanderlust Rambling Club, formed in 1932 in Grimsby by a certain Mr Neville Cole. Our attention was drawn to the club's existence by a little moss-covered plaque hidden amongst a shady grove dappled with primroses. The route is almost 60 miles long, making use of the Jurassic Ridge formation to provide good viewpoints along the way. However, this route is now partly abandoned due to the A15 which cuts the pathway off, with no pedestrian footbridge to provide safety. In order to safely divert the route, an adjacent landowner must give permission for access. However, according to informed interviewees, the landowner is based overseas and it is proving difficult to contact. What can we make of the complexities of land access and rights and the development of a controversial road scheme? As we reflect on in Chap. 3, the use of wetland spaces for health and wellbeing is often a challenge for a wide variety of reasons.

Letting the Nev Cole Way fall into lapsed use, rather than developing it as a facet of local history and heritage closely linked with the new wetland site, is also reflected in the current low visibility of the wartime use of Alkborough Flats. During the Second World War over one-third of Bomber Command was based in Lincolnshire and the area now known as Alkborough Flats was used for aerial target practice by fighter planes (Wood 2016).

Structures remain at Alkborough Flats today, such as bomb surveying posts, look-out posts (see Image 2.12) and the foundations for search-lights. The look-out post shown above sits on the escarpment and is just one element of a chain of defensive buildings including anti-tank traps and pill boxes. The level of decay to this look-out post speaks volumes; this heritage is one not currently valued. The coastal realignment is an

Image 2.12 Look-out post at Alkborough Flats. (Credit: Geography.co.uk)

aesthetic re-presentation of this space, and clearly its military past sits somewhat uncomfortably with this new vision. An archaeologist working on the site during the realignment works confirmed a number of wartime ordnance artefacts were removed from the site, though there is no on-site information about wartime engagement. In many ways this information remains in oral history and artworks (see Image 2.13) and is kept alive by local historians.

The current site seems to not have emphasised the part that the geography played in the war effort. Instead, other heritage uses seem more compelling in the creation of a narrative about the site within which nature plays a more dominant role. The leading example on this site is Julian's Bower (see Image 2.14)—a medieval turf labyrinth which sits on the escarpment above the Flats, adjacent to Alkborough church.

Image 2.13 Werner 162. (Credit: Maxim Peter Griffin)

Night near the tank traps -

brilliant flurries of practice fire

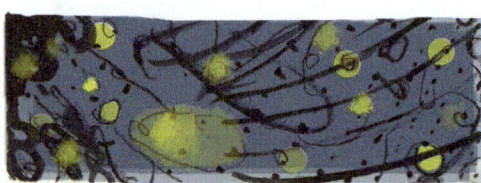

at the jaws of mighty Humber

Image 2.14 Julian's Bower, Alkborough Flats' turf maze, February 2018. (Credit: Mary Gearey)

The labyrinth, also referred to locally as a turf maze, is replicated in the entryway to the church. As a result the church and village utilise Julian's Bower for social events such as marking Easter and for national celebrations such as the beacon lighting ceremony to mark Queen Elizabeth II's 90th birthday (BBC News 2016). Its importance stands as not only a scheduled ancient monument, but to recognise the importance of this waterscape to humans over time into antiquity. It is thought the current labyrinth stands on the site of a Roman maze, with local archaeological finds on the Flats verifying Roman influence and occupation of this area, and that they were used for marking social practices such as courting (Stukeley 1776; Higgins 2018).

The importance of this landscape when we consider global movements of populations and cultural practice is also noted by local historians who lead group tours to explore the Pilgrim heritage of the area. Close to the Flats is a raised platform earthwork huddled within a copse of trees. This earthwork, a human-made construct embedded in the landscape, has been identified as Saxon era in origin, but as with Julian's Bower, could be Romanesque. It has the moniker 'Countess Close', referring to a local Saxon queen. More recently, in the seventeenth century, it was used by

Pilgrim preachers to host clandestine meetings to discuss separation from the Church of England, which, though also protestant in belief, they felt was too Catholic and ostentatious (Moore and Moore 2007). These secret meetings in spaces such as Countess Close again attends to the changing ways in which the waterscapes of Alkborough Flats were deemed to be out of the way and too remote for the prying eyes of the capital. Some pilgrims from the Alkborough area were the original Mayflower pilgrims who sailed from Plymouth in September 1620 in search of religious freedom. The historical legacy of the case study sites is most evidently found in our Somerset Levels study.

The Somerset Levels: Shapwick Heath and Westhay Moor

The Somerset Levels are a vast network of intersecting wetlands, sitting within the drainage basins of the rivers Brue, Tone, Parrett and Axe in the South-West of England. These collective river valleys sit within a vast flat topography of 250 square miles (Williams 1970: 2), bordered by the Mendip Hills to the north, the Quantocks to the south and the smaller range of Polden Hills separating the river systems roughly in the Levels' midsection. The Polden Hills not only influence the drainage regime of these connected wetlands through separating the north and south aspects of the Levels, but they also influence rainfall and wind direction, exerting subtle differences between the two halves. When the area experienced heavy winter flooding in 2013/2014, it was the area south of the Polden Hills that bore the brunt of the pluvial and fluvial storm water. Given the influence of southwesterly winds coming in from the Atlantic, this 'West Country' section of the UK is prone to higher than average levels of rainfall when compared to more easterly parts of the UK (MetOffice 2019). The low-lying topography and clay/peat soil structure, coupled with the abundant rainfall and tidal influence of the four riverine systems, make this a rich wetland landscape (Image 2.15).

As with other 'Levels' in the UK (Pevensey Levels in East Sussex, Seasalter Levels in Kent, Humberhead Levels, spanning Lincolnshire and South Yorkshire amongst others also across Wales, Scotland and Northern Ireland), these typify lowland wetlands. The term Levels is both descriptive and vernacular, as other flat marshland areas within the UK are termed fens, carrs and broads—and all describe very similar 'topographic and hydrographic' terrains (Williams 1970: 2). Mitsch and Gosselink (2015)

Image 2.15 Shapwick Heath's rich diversity. (Credit: Adriana Ford)

describe these types of wetland as being either composed of peat or peat forming. In other words, within these spaces, there is a prevalence of organic material and reliable levels and frequency of water inundation via rainfall, surface and groundwater recharging.

These waterscapes typify those wetland environments which have strong cultural resonances as 'classic' wetlands. Their aesthetic appeal accords with data from the project interviews where participants described what they thought were 'real' or 'proper' wetlands. They typically were vast areas of water and reed bank, with bird hides enabling bird watching and wildlife photography. Alongside this are concepts of antiquity—that these are landscapes barely touched by human agency and are spaces where humans have lived intimately with the landscape in pre-modern incarnations. This has always suggested forms of hidden abundance, such as eel fishing, reed cropping for baskets, traps and thatching, wildfowling, willow coppicing and peat extraction, as well as collecting flowers, herbs and other plants for traditional medicine. In other global wetland systems, this provisioning would include large-scale agro-industries such as rice production or cranberry harvesting (Mitsch and Gosselink 2015). Increasingly, as we will explore below and in other chapters in this book, ecotourism and forms of leisure activities associated with these environments provide new income streams for communities who live within the wetlands. This is particularly true for the Somerset Levels that have benefitted from additional forms of tourism brought in by interest in the Glastonbury Music Festival (https://www.glastonburyfestivals.co.uk/) and associated arts led activities in the area (see Fig. 2.5).

The case study fieldwork focused on two wetlands within this waterscape network: Shapwick Heath and Westhay Moor. In their most recent incarnation (in geological terms), over the last eight hundred years or so, these sites have been artificially drained and used for agriculture, livestock and peat extraction. Though there is not a linear graph with regard to drainage, inasmuch as drainage activities stopped and started throughout this period, we have, since the early 1990s, seen a different approach taken on the case study sites: to end large-scale hydraulic draining practices and return to something more prosaic, based on a heuristic assessment of acceptable water levels in the area. However, as will be explored, it is whose heuristics that determine the current water management regime in these spaces. For now, before taking a closer look at the two wetlands and their local context, it is useful to

Fig. 2.5 Shapwick Heath and Westhay Moor in relation to Glastonbury

emphasise the importance of the term 'drainage' within the Somerset Levels, as, to a great extent, the draining of marshlands across the UK and Northern Europe from the twelfth century onwards can be closely associated with the twin practices of nation building and incipient capitalist forms of political economy.

The Somerset Levels region, like so much of lowland marshes within the UK, has been dramatically shaped over many historical epochs through human agency. Of all of our three case studies, the Shapwick Heath and Westhay Moor examples highlight changing attitudes towards wetlands: from drainage on a local scale in the medieval era, through to the era of national hydraulic drainage practices masterminded by Dutch engineers, to local jurisdictional governance via the Commission of Sewers. Then from the 1930 Land Drainage Act until the early 1990s, via the Internal Drainage Boards, we had a steady-state approach of water resources management which has changed since 1989 with the privatisation of the English water supply sector which has impacted the management of water resources more widely, as discussed below. Drainage and wetlands are intimately connected; the 'drain and reclaim' approach has dominated our English wetland environments for hundreds of years—a global paradigm of land management which has been slow to usurp.

Drainage has, until very recently, been pitched as the 'solution' for wetlands (Verhoeven and Setter 2009). From the 'High' Middle Ages of the twelfth century onwards, marshlands were viewed as inhibiting development, both in terms of economic and social progress (Schneider 1990). These landscapes were seen to be held under the control of nature, thwarting human agency and threatening food security. As global markets were beginning to open up through seaborne and land travel routes, land became prized for its productivity; satisfying domestic demand was only part of this emergent international trade network. Wetlands were seen as wastelands. Neil Smith's 1984 seminal work outlined the specific ways in which nature has been co-opted into an increasingly globalised capitalist paradigm. His key text, *Uneven Development: Nature, Capital, and the Production of Space*, is vital to our analysis of the ways in which wetland ecosystems have been (mis)used and (mis)valued in the modern era. Smith utilises a Marxist critique to alert us to the ways in which our human relationships with landscape, ergo nature, have been shaped through historical epochs within which episodes of uneven economic development have supported forms of political economy orientated around private ownership of land and its attendant resources. As we see in Chap. 5, culture plays a significant role in the development of particular kinds of human-nature narratives which give credence to market capitalism. Literature over time, along with folk stories and other oral traditions, have in many differing ways supported these narratives of useful and desirable space, casting marshland in particular as useless, unwanted and redundant space. As we will see, these depictions both support the hegemonic perspectives of wetlands as wastelands and spur counter-hegemonic practices which utilise these perspectives to keep these spaces as othering places for delinquent activities.

During the early modern era, low economic activity on wetlands was viewed as endemic; without alternative options, these fens and boggy terrains would remain as stagnant places—in all senses of the word. This stagnancy extended to the communities that lived on the marshes, who were viewed as insular and unwilling to let go of traditional livelihoods and cultural practices. As we will see, drainage was closely linked with modernity, the flow of water shaping social and economic relations and development in these terrains. Sultana (2013: 349) has described these formative processes of reshaping landscape as 'the political ecologies of development' through which the 'liminalities of water technologies and vagaries of heterogeneous nature/water, brings to the fore the contested and contradictory processes of development itself'.

Attempts to undertake drainage reforms in the medieval period in Somerset have been described by Williams as 'rudimentary' (1963: 166). The piecemeal nature of attempts by landowners to reclaim this land through water removal via dykes and rhynes, artificial channels to speed the removal of water from the surface and root zone, is evidenced by archaeological records which show periods of use and abandonment of these techniques (ibid: 164). The Somerset experience reflects the historical trend at this time across England. The medieval period is a vast span of historical time, with Early, High and Late Middle Ages spanning from the ninth to the sixteenth centuries (Turner and Young 2007). These patterns of technological update and lapse are indicative of pre-capitalist forms of land use—smaller fragmentary pieces of land, owned and used by families, collectives and communities through a mix of traditional land rights and commoning.

Of crucial importance is a consideration of the transition from the pre-modern to the modern. Land use and drainage at this time are closely associated with the expansion of Christian perspectives, with a targeted set of socio-cultural practices to move away from pre-Christian beliefs in which certain land forms were populated with malevolent spirits. Roymans (1995: 3) details how 'a sacral, cosmologically-embedded ordering of space' forming pagan belief systems had to be broken in order to reclaim untouched marginal land such as the moors, heaths and bogs of wetlands. As we will outline in Chap. 5, the spirits which were believed to inhabit these spaces, including will-o'-the-wisps, witches and subterranean dwarves, were thought to bring bad luck to humans, their animals and crops. It was also believed unwise to tamper with these liminal zones.

As a result these spaces were often used for burials and othering practices. Bender (2002) notes how these 'peasant' or subaltern responses to land reclamation were possibly framed less around superstitions but used these as a smokescreen to really defend against the encroachment of state practices in these territories. Claiming these spaces as workable/worked land was then a propagandist tool for Christian expansion and a tool for economic development and forms of new social ordering. Drainage then can be viewed as an element of complex theocratic empire building during this stage of English history. Yet these social upheavals are still in a dynamic relationship with exogenous influences. Bailey (1989) details how land use fluctuated in tandem with demographic changes, particularly in response to a shrinking population in line with various epidemics and socially transmitted diseases including the Black Death, smallpox and leprosy. His work focuses on the uptake of marginal land as populations rise; and we can consider here the ways in which wetlands can fall within this category.

From the twelfth century onwards, reaching a peak in the seventeenth and eighteenth centuries, the English landscape was transformed through hundreds of Enclosure Acts, annexing fields and meadows for agricultural use. In tandem with this nation-wide move to individualise land ownership were actions to drain and reclaim these varied spaces, of which wetlands form part of the matrix. The 'Court of Sewers' were established by King Henry VIII in 1531 by the 'Act of Sewers' (https://www.livinglevels.org.uk/stories/2018/12/10/court-of-sewers) to oversee these land reforms, starting in coastal marshlands. Sewers were the vernacular term for constructed drains, and not the wastewater pipe network we associate with the word in current practice.

The Somerset Levels then were classic coastal wetland complexes co-opted into this regime. These 'Court of Sewers' were responsible for sea defences, listened to appeals and could enforce the drainage responsibilities placed on local landowners. These Courts were transformed into 'Commissions of Sewers' by the sixteenth century. In line with the beliefs of those in power Enlightenment thinking was foundational in the ruling of these Commissions; they used rationality and 'science' as the purported means to come to just and rational conclusions for the public good. These Commissions were constituted by local jurors, foresworn to base impartial judgements on the evidence presented to them with regard to new engineering works on drainage or for conflict resolution. These jurisdictional bodies populated other countries too, in particular the Netherlands, within regions where drainage proved to be of high social, economic and political importance (Morgan 2017). These organisations flourished during the sixteenth and seventeenth centuries as land management increasingly came under state control within a strategy of taxation and nation building. Historical records suggest that although these courts were weak and fragmentary due to their disparate catchment allocation, and this is particularly the case in the Somerset Levels (Williams 1963), they are significant because they ushered in a process of text rather than oral record as precedent, enabling the educated social classes to further embed their dominance over lower social orders (Morgan 2017). However, Holmes (1985) suggests that these Commissions enabled those in marginal peripheral spaces, such as the East Anglian fens, to learn about judicial and political processes that were happening in the urban central spaces and so to develop capacity around modern processes of decision making. We can see then that current critical theorists' observations that water, society and power building are inextricably linked can be identified in the switch towards modernity (Swyngedouw 2004).

The Land Drainage Act of 1930 led to the ending of the Commissions and the centralisation of policy decisions around drainage issues. Replacing the Commissions were the Internal Drainage Boards (IDBs), many of which are still operational today. IDBs are tasked with operating as statutory bodies to oversee drainage issues and are generally comprised of a range of stakeholders including riparian landowners, tenant farmers, local and parish councillors. They are organised around river catchments, rather than political or administrative jurisdictions, and so in many ways have acted as the precursor for modern water management technical modes of operation, particularly in England (Werritty 2006).

For wetlands this centralisation of water resources management marks an acute rise in the use of industrial techniques to develop arable land for agriculture. The escalation in the scale and temporality of social change from the 1930s onwards has been well documented by social and cultural geographers (Harvey 2010; Massey 2004; Rose 1993; Soja 1989) and by ecologists seeking to raise awareness of wetland decline (Joyce 2012). For our consideration of wetlands, it is pertinent to finish here by suggesting that our sense of self is closely linked with the landscapes we live and work amongst. Rapid changes to our land and waterscape can unsettle this sense of self: 'different configurations of water "act" as the "ground" for various forms of identity' (Matless 2000: 143). The work of cultural geographers such as Matless and Dorren massey help us to read changes to English wetlands over different socio-historical periods as also impacting on our personhood. Our relationships with landscape are always in flux; and these changes, in turn, shape our embodied sense of self. In this way our interior lives are closely connected to the changes in the landscapes we inhabit (Drenthen 2009), and our sense of dis-ease can be linked with rapid socio-natural technological change. As much as we feel this presciently now with climate change impacts creating uncertainty and unsettling our sense of the known, we can also have empathy with other humans in past environments who felt this sense of disorientation too in respect to dramatic anthropogenically driven landscape change. We explore this in a different way in the following chapter as we turn to another fundamental aspect of human-landscape connectivity: play.

References

Bailey, M. (1989). The concept of the margin in the medieval English economy. *Economic History Review*, (1), 17.

Barker, A., Booth, N., Churchill, D. and Crawford, A. (2017) The Future Prospects of Urban Public Parks: findings – informing change. Available at: https://futureofparks.leeds.ac.uk/wp-content/uploads/sites/26/2017/07/Job38853Future-of-Parks-Findings-Report.pdf. Accessed: 01/06/19.

BBC News. (2016) Queen's 90th birthday: Beacons lit amid UK celebrations 21st April 2016 (https://www.bbc.co.uk/news/uk-36093306). Accessed on 11th May 2019.

Bender, B. (2002). Contested landscapes: medieval to present day. In V. Buchli (Ed.), *The material culture reader* (pp. 141–174). Oxford: Berg.

Bishop. K. (1991) Community forests: implementing the concept. The Planner 77 6–10.

Cook, B., et al. (2016). Competing paradigms of good management in the Scottish/English borderlands. *Disaster prevention and management: an international journal, 25*(3), 314–328.

Dempsey, N., Burton, M., & Selin, J. (2016). Contracting out parks and roads maintenance in England. *International Journal of Public Sector Management, 29*(5), 441–456.

Drax. (2017) Sustainability. https://www.drax.com/sustainability/people/#targetText=Job%20creation&targetText=We%20directly%20employed%20a%20total,to%20the%20economy%20in%202017. As accessed at 6th September 2019.

Drenthen, M. (2009). Ecological restoration and place attachment: emplacing non-places? *Environmental Values, 18*(3), 285–312.

Forest of Marston Vale Trust (2008) Our Story. https://www.marstonvale.org/our-story. As accessed 19th May 2019.

Gray, M., & Barford, A. (2018). The depths of the cuts: the uneven geography of local government austerity. *Cambridge journal of regions, economy and society, 11*(3), 541–563.

Harrabin, M. (2019) Climate change: England flood planners 'must prepare for worst' 9th May 2019. Online news report: (https://www.bbc.co.uk/news/science-environment-48206325). Accessed on 17th June 2019.

Haraway, D. (2015). Anthropocene, capitalocene, plantationocene, chthulucene: Making kin. *Environmental humanities, 6*(1), 159–165.

Harvey, D. (2010). *Social justice and the city* (Vol. 1). Georgia: University of Georgia Press.

Heritage Lottery Fund. (2014). *State of UK public parks 2014 – Renaissance to risk?* London: Heritage Lottery Fund.

Higgins, C. (2018). *Red Thread: On Mazes and Labyrinths*. London: Random House.

Holmes, C. (1985). Drainers and fenmen: the problem of popular political consciousness in the seventeenth century. In A. Fletcher & J. Stevenson (Eds.), *Order and disorder in early modern England* (pp. 166–195). Cambridge: Cambridge University Press.

Holstead, K. L., et al. (2017). Natural flood management from the farmer's perspective: criteria that affect uptake. *Journal of Flood Risk Management, 10*(2), 205–218.

Humber Nature Partnership (2020). See https://www.humbernature.co.uk/estuary/. Accessed on 7th July 2020.

Ingold, T. (1993). The temporality of the landscape. *World archaeology, 25*(2), 152–174.

JBA Consulting file:///C:/Users/Mcag10/Downloads/CaseStudy1Alkborough FlatsTidalDefenceScheme-FD2635%20(1).pdf

Joyce, C. (2012). Preface: Wetland services and management. *Hydrobiologia, 692*(1), 1–3.

Lawrence, A., & Ambrose-Oji, B. (2015). Beauty, friends, power, money: navigating the impacts of community woodlands. *The Geographical Journal, 181*(3), 268–279.

Massey, D. (2004). Geographies of responsibility. *Geografiska Annaler: Series B, Human Geography, 86*(1), 5–18.

Matless, D. (2000). Action and noise over a hundred years: the making of a nature region. *Body & Society, 6*(3–4), 141–165.
Ministry of Housing, Community and Local Government. (2019) The Oxford-Cambridge Arc Government ambition and joint declaration between Government and local partners. Stationary Office. https://assets.publishing.service.gov.uk/government/uploads/system/uploads/attachment_data/file/799993/OxCam_Arc_Ambition.pdf as accessed 011019. As accessed on 23rd June 2019.
Mell, I. (2018). Establishing the costs of poor green space management: mistrust, financing and future development options in the UK. *People, Place and Policy, 12*(2), 137–157.
MetOffice. (2019) UK regional climates. https://www.metoffice.gov.uk/research/climate/maps-and-data/regional-climates/index. As accessed on 3rd May 2019.
Mitsch, W. J., & Gosselink, J. G. (2015). *Wetlands* (5th ed.). New York: John Wiley.
Moore, S. H., & Moore, S. M. (2007). *Pilgrims: New World settlers & the call of home*. Yale, Yale University Press.
Morgan, J. E. (2017). The micro-politics of water management in early modern England: regulation and representation in Commissions of Sewers. *Environment and History, 23*(3), 409–430.
Payne, J. (2017). My Place: Scunthorpe. *Teaching Geography, 42*(2), 75–76.
Rose, G. (1993). *Feminism & geography: The limits of geographical knowledge*. Minneapolis: University of Minnesota Press.
Roymans, N. (1995). The cultural biography of urnfields and the long-term history of a mythical landscape. *Archaeological dialogues, 2*(1), 2–24.
Schneider, J. (1990). Spirits and the spirit of capitalism. In E. Badone (Ed.), *Religious orthodoxy and popular faith in European society* (pp. 181–220). Princeton: Princeton University Press.
Smith, N. (2010). *Uneven development: Nature, capital, and the production of space*. Georgia: University of Georgia Press.
Soja, E. W. (1989). *Postmodern geographies: The reassertion of space in critical social theory*. London: Verso.
Stukeley, W. (1776). *Itinerarium Curiosum; Or, An Account of the Antiquities, and Remarkable Curiosities in Nature Or Art, Observed in Travels Through Great Britain: Centuria I* (Vol. 1). Cambridge: Baker and Leigh.
Sultana, F., 2013. Water, technology, and development: transformations of development technonatures in changing waterscapes. Environment and Planning D: *Society and Space, 31*(2), 337–353.
Swyngedouw, E. (2004). *Social power and the urbanization of water: flows of power*. Oxford: Oxford University Press.
Turner, S., & Young, R. (2007). Concealed communities: the people at the margins. *International Journal of Historical Archaeology, 11*(4), 297–303.

Verhoeven, J. T., & Setter, T. L. (2009). Agricultural use of wetlands: opportunities and limitations. *Annals of botany, 105*(1), 155–163.
Werritty, A. (2006). Sustainable flood management: oxymoron or new paradigm? *Area, 38*(1), 16–23.
Williams, M. (1963). The draining and reclamation of the Somerset Levels, 1770–1833. *Transactions and Papers (Institute of British Geographers)*, (33), 163–179.
Williams, M. (1970). *The draining of the Somerset Levels*. Cambridge: Cambridge University Press.
Wittering 2017. See: https://www.lincolnshirelife.co.uk/posts/view/alkborough-ancient-footprints
Wood, L. (2016). *Little Book of Lincolnshire*. Cheltenham, UK: The History Press.

Websites

https://www.bedsfire.gov.uk/About/Bedfordshire-Fire-and-Rescue-Service.aspx
https://www.bbc.co.uk/news/uk-36093306
https://bedsarchives.bedford.gov.uk/CommunityArchives/Bedford/TheNewStoneAgeInBedford.aspx. Accessed at 01102019
https://www.glastonburyfestivals.co.uk/
https://www.heureka.clara.net/lincolnshire/alkborough.htm
https://www.livinglevels.org.uk/stories/2018/12/10/court-of-sewershttps://www.marstonvale.org/our-story

CHAPTER 3

Wetlands as Ludic Spaces: Play, Recreation, Rejuvenation, In/Exclusion

Abstract This chapter explores the ways in which English wetlands have been repositioned as 'ludic' spaces—recreational places within which to enhance individual and collective wellbeing. This chapter draws on fieldwork undertaken within the project to evidence what types of ludic activities are gained on different wetland sites and what this means for human-wetland connectivity within these waterscapes. Walking and cycling in wetlands, admiring the spectacle of autumn starling murmurations, joining organised health and fitness activities on site—all are promoted as rejuvenating, restorative practices, offering a wide range of personal benefits. Focusing on three key aspects—on-site amenities, access and safety—this chapter will outline the different ways actors within these three case study sites interface with and experience conflicts, barriers and opportunities for ludic play. Underpinning this is a desire to interrogate ideas and management practices around 'natural' wetland environments and ludic play, exploring delinquency, self-expression and othering practices in these ecosystems.

Keywords 'Ludic' • Recreational • Wellbeing • Human-wetland • Waterscapes • Murmurations • Restorative • Delinquency • Self-expression • Othering practices

This chapter explores the ways in which English wetlands have been purposively repositioned as 'ludic' spaces—recreational places within which to spend time alone or with loved ones, spaces in which to enhance one's own wellbeing. Environmental tourism, whether walking and cycling in wetlands, viewing murmurations of autumn starlings, and joining health and fitness activities on site, all are promoted as rejuvenating, offering a wide range of personal benefits. As a result visitor numbers are still seen as one of the key indicators for how well used and well regarded different wetland spaces are with the general public. Yet there is little research to interrogate the different ways wetlands are enjoyed in varieties of ludic play, with all the richness, and challenges, which that may bring to wetland managers and those tasked with improving social diversity in these spaces of protected biodiversity. This chapter draws on fieldwork undertaken within the project to evidence what types of ludic activities are performed within different wetland sites, what this means for human connectivity with these spaces and wider considerations of what constitutes inclusive 'play' (DEFRA 2018). Explored across a number of wetland types within the case study sites (visitor centre orientated, country park or less managed space with few amenities), the chapter explores the deliberate reinvention of wetlands as ludic spaces—and how these human-nature prescriptivities serve unwittingly to exclude other players. There is now a large body of writing on the power relations, strategies, rules, norms, practices and representations that exclude certain social groups from leisure spaces including teenagers, ethnic minorities, homeless people, migrants and older people (Church and Coles 2006; Oncescu and Neufeld 2019). Indeed, the excluding tendencies of the national landscape protected areas that cover 24.5% of England have been a key focus of the ongoing government review into these areas, and the interim report was clear-cut in stating that when it came to being welcoming and inclusive to all groups in society, 'many landscape bodies have not moved smartly enough to reflect this changing society, and in some cases show little desire to do so' (DEFRA 2018: 2). In the UK the inclusive and exclusive dimensions of leisure and play spaces have resulted in significant conflicts and disputes being a long-standing feature of some wetlands and waterscapes that shape what forms of play and use are permitted. In many locations water users such as anglers, canoeists and boaters come into conflicts with each other and also with waterside users such as walkers, birders and cyclists (Church and Ravenscroft 2011). These conflicts may involve small irritations linked to noise and having to share water space, but in some cases physical conflicts have arisen and landowners have taken legal action to exclude certain users from rivers and wetland spaces. In this chapter we identify how certain forms of play involving alcohol, drugs and sex, that conflict with the goals of wetland managers and other users, have been managed by legal by-laws but still take place usually hidden from the other users who might be negatively affected. A number of long-standing perspectives exist to explain conflicts linked to outdoor leisure (Jacob and Schreyer 1980), and Carothers et al. (2001) argue that two main categories of conflict exist. The first is 'interpersonal' conflicts that occur because of goal interference when one user's actions harm the goals other users are seeking. In our three wetlands, many of the birders using the hides had worked out how to adjust their visiting times to avoid school and family groups. The second category of conflicts is defined as 'social value' conflicts and arises from differing views on the social acceptability of particular users (Carothers et al. 2001). Social value conflicts can arise without people needing to be in physical proximity to one another, and the social construction of space and its role in influencing who feels welcome is a key factor in determining the type of conflicts that arise (King and Church 2015).

In this chapter we illustrate how our case study wetland spaces are physically designed and promoted in different forms of media often to create social norms and values that whilst seemingly inclusive tend to prioritise younger children and families as desired visitors. In some of these locations, however, the management and social construction of the wetland accommodate users who in other locations have come into conflict. Slalom canoeing often conflicts with anglers who want exclusive use of fast-moving waters and local residents who object to large numbers of visitors at key locations (Church and Ravenscroft 2011). In PCP in Bedford, slalom canoeing has been designed into the wetland to create a managed space separate from other water users but also to encourage other visitors to spend time watching the canoeists. This is a reminder that there are limitations to seeking to understand conflicts as arising solely from interpersonal goal interference and social values. Often the distinction between the two is blurred especially where physical conflicts occur over both values and goals (Ravenscroft et al. 2002), and importantly power relations come 'into play' in that whose play is permitted is determined by property rights and legal arrangements that the state and landowners can use to include and exclude certain users. In other water and wetland locations, the nature of play is determined by the ability of landowners and certain leisure users, often anglers with acquired property rights, to mobilise their power through the law and property to include and exclude certain users (Church et al. 2013). The managers of our case study wetlands, as this chapter shows, all have a strong commitment to attracting users and being inclusive. The previous research alerts us, however, that we have to be mindful that whilst wetlands allow distinct cultural experiences and excite the imagination, they can also be spaces of conflicts between users, owners and managers, resulting in inclusion and exclusion based on property rights, the law, goals and values. Also certain groups may feel excluded by features of the inclusive actions that construct wetland spaces.

Curiosity, Learning and the Ludic

One of the seminal texts regarding the human drive to play is that of the Dutch sociologist Johan Huizinga in his 1950 text *Homo Ludens*. Huizinga argued that play was a necessary component of learning through creative engagement with new subject matter and, further, that the human impulse to acquire knowledge is driven by curiosity. Active learning can take place through play, in all its forms and in any type of environment. Rodriguez (2006: 1605) goes further to suggest: 'the fundamental motive of play is the experience that it affords'. If we consider affordance to be a crucial phenomenological aspect of human and environment interactions, then we can see that the types of human-environment experiences, performances, co-responses—define them as you will—that we find in wetlands and which are fuelled by curiosity and imagination can be understood as ludic play in support of learning, whether self-directed or open to the world.

The interactions between learning, play and nature are also recognised in a more instrumental way by ecosystem services assessments. Many of these assessments seek to categorise and measure the benefits humans gain from nature into three broad categories of services provisioning (e.g. food and fibre), regulating (e.g. climate regulation) and cultural. The latter has caused much debate over conceptualisation, definition and terminology (Fish et al. 2016). Perhaps reflecting the complexities of the term culture, many national ecosystem assessments have used up to seven different cultural services subcategories to include tourism, leisure, aesthetics, education, identity, sense of place and spirituality. Play is primarily considered in assessments under leisure with education often focussing on formal learning. The IPBES (2019) global ecosystem assessment emphasises the term 'nature's contributions to people' rather than ecosystem services, partly as it is seen as a more inclusive term that incorporates ecosystem services into a wider conceptualisation. Tellingly, the term cultural was replaced by the IPBES with 'non-material' (Díaz et al. 2018). In IPBES (2019) the number of non-material (cultural) categories was reduced to four: learning and inspiration, physical and psychological experiences, supporting identity and maintenance of options, the latter being concerned with future benefits from nature. It was also recognised that activities such as play and leisure cut across these categories as they contribute to learning, experience and identity (Díaz et al. 2018).

Whilst ecosystem services approaches provide useful categorisations for understanding the interactions between learning, play and nature, they can over-emphasise instrumental perspectives that see play as purposeful and goal oriented. By contrast, Woodyer (2012) considers how limited portrayals of play frame playing as a child orientated activity, used by children to rehearse 'real' life. Woodyer contends that this approach both denigrates the immersive and self-defined subjectivity of play, as a truly anarchic, unbounded free expression of self, and casts adults as sitting outside these forms of performativity. This is important when we wish to consider what Woodyer calls 'a politics of playing' (ibid: 317) alongside a consideration of wetlands as wellbeing spaces. Some forms of play are more acceptable than others. As we see from the interviews and case study observations, detailed 'play' means different things to different sociocultural groupings and this has significant implications for wetland managers if they wish to present them as inclusive wellbeing spaces.

PLAY: RECREATION, REJUVENATION, RESTORATION

I don't care what any of the rest of you say, there's nothing better than a May morning when the apple's in blossom and there's a thickness and you walk around and you almost step on a deer and you hear the cuckoo calling and all that, and that is what the Moors and Levels are about, first thing in the morning. Yeah, everything, yeah, everything, you go quiet in there and you feel you're in heaven really. (Farmer, Somerset Levels, female, 70s)

Play is a troubled concept in these neoliberal times. What value is there in play—in taking time to do nothing, to be in the moment, frivolous and undirected? To play is not to work. To work is to be effective, productive, remunerative. Play is what happens outside of work; it restores you so that you can return to work refreshed. Play is a component of our leisure time—but only an element. It sits in a dialectical relationship with work. Play in a non-waged framing is idling, lounging, *dossing about.*

The old, the young and those unencumbered by work are those who have time to play; moreover they have the social and cultural legitimacy to play, but not necessarily the resources to play. For those who work, playtime is syncopated; play is squeezed into the margins of everyday life. Play becomes a serious business, requiring attention and forms of specialist knowledge.

So what does it mean to play in wetland spaces—these 'quaking' (Giblett 2009: 1) zones where earth and water meet? What playmates have the guile to cross treacherous bogs and insect laden heaths (see Image 3.1)? The 'bog of eternal stench' from the 1984 film *Labyrinth* is an iconic representation of paludal, marshy, wetland environments. Mystery and fear reside here in abundance. Here the rational world is left behind; the wetland is a transitional zone where our primal fears and childish anxieties are let loose (Image 3.2). Is there something syntactically, anagrammatically and playfully pertinent about the words ludic and lucid in these liminal spaces where large fen skies and head height bulrushes throw out our sense of perspective as we perambulate around these waterscapes?

Image 3.1 The bog of eternal stench: still from the film *Labyrinth*

Image 3.2 Labyrinth, Priory Country Park, Bedford, January 2019. (Credit: Phil Shirley)

Played Out: Family Leisure Time in Wetlands

We start our quest by referring to what Blackshaw (2010: 141) has termed 'liquid leisure'—an understanding of leisure as a 'hermeneutical exercise… that provides us with our own unique view of the world'. Suggesting as it does a distancing from the old work/play dualism (Chubb and Chubb 1981; Gershuny 2000), with its Marxist implication of a developmental shift from meaningless work to meaningful playtime (Opaschowski 1998), Blackshaw offers a new perspective on the ways in which people understand and deploy time (Rojek 2010a, b; Roberts 2011; Veal 2012), by heralding a form of 'leisure society' characterised by a holistic approach to the social, psychological and financial practices associated with work, leisure and play (see Ravenscroft and Gilchrist 2009, for a description of the working society of leisure). Blackshaw thus offers a way of thinking about leisure that moves away from *activity* as the defining practice—of work and leisure—towards *activities* as instrumental contributions to the performance of self and family through leisure and play.

This idea draws heavily on Aristotelian philosophy (see De Grazia 1964; Hemingway 1988) that leisure should be 'an arena for the active development of the individual as a member of the community' (Mair 2002: 215). This is very much emphasised in our wetland sites, where people—individuals and families—are encouraged to express their selves in ways that challenge conventional social and cultural boundaries. Indeed, many wetlands and other public spaces are increasingly packaged to encourage the deployment of what are essentially inconsequential and regulated leisure practices in ways that offer the possibility of engaging in meaningful and transformative civic environmental projects (Ravenscroft and Taylor 2009):

> It is my contention that in the liquid modern world we live in, which is founded first and foremost on freedom, leisure moves steadily into its position as the principal driving force underpinning the human goal of satisfying our hunger for meaning and our thirst for giving our lives a purpose. This is the job that leisure was always cut out for, since it is that distinct realm of human activity which perhaps more than any other provides us with the thrill of the search for something and the exhilaration of its discovery. (Blackshaw 2010: 120)

This, then, is our contention about the opening of wetlands and other such sites to public access for recreation and play. While being inconsequential practices in themselves, taking place in what are often relatively mundane locations, these leisure activities bind people together in shared—civic—understandings of what it is to be 'in place' in wetland sites. For Mair (2002), these wetlands thus become 'civic commons' that are quite possibly unbounded in people's imaginations, while in terms of inclusion and exclusion often remaining tightly regulated in terms of permitted access and activities (see Gearey et al. 2019). These ludic spaces thus offer a challenge to conventional views about practices associated with leisure as self-serving, privatised and consumerist (Rojek 2001; 2010a, b). Indeed, as Sumner and Mair (2017) argue, leisure can be understood as a form of life good that is deployed as part of a moral career in which the focus is not so much on income or wealth generation, but on experiences—often in ludic spaces—that bring new purpose to people's lives through a conjunction between activity, purpose and space(s). They go on to suggest that, increasingly, the activities are part of what they refer to as 'sustainable leisure' understood as a range of activities that contribute to improving the civic commons. For them, activities that build up the civic commons make society more sustainable—underpinning the significance of wetland sites.

Family leisure time, as part play, part embedded non-waged labour, thus forms a substantive part of the rationale for encouraging wider wetland use and access. It is, of course, also a distinct type of income generation category. When we cast around to look at 'play' on wetlands, we see vastly differential forms of leisure activities enacted. Within the Country Parks of the Bedfordshire case study sites, there are spaces for young family play, small playgrounds in PCP and a wooden labyrinth to hide within from adult sight, picnic tables close to the lakes to feed the birds, flat paths for pushing prams and toddler wayfinding, spaces for cycling and jogging together, information boards for on-site learning, exhibition space at MCP for displays and community groups to share information. At PCP there is a marina with opportunities to learn to sail, windsurf and paddle board along with a fast water channel for slaloming racing in canoes.

The bird hides across all three case study sites offer families interactive learning via daily sighting charts, bird identification notices and through online updates from each hide location from the site rangers (Fig. 3.1).

These are also teaching spaces for pedagogic practice, with classroom activities organised through interactive play involving observation, interpretation and embodied learning, that is, getting muddy and wet. These activities also help reconnect families with nature and with each other:

> We did a, a dads and daughter's day at Priory, the girls did canoeing and paddle boarding and then they had a nature trail where they had to, erm, run around the country park finding photos of things and they had to mark where the photos were on, on a map, erm, and we got incredible feedback from the dads because they said that they'd have never done this with their daughters otherwise. (Guiding Association leader, Bedfordshire)

Fig. 3.1 Guide to Shapwick Heath. (Credit: field-studies-council.org)

The Avalon Marshes Centre in Somerset also provides visual history timeline information, presented in a format which is comprehensible to school-age children, connecting the present-day space to the Neolithic practices found on the site. The 'Sweet Track', a Neolithic raised wooden trackway, over 5000 years old, discovered on Shapwick Heath during the engineering excavation work in 1970 is an exemplar of this. A modern reconstruction of some of the trackways now exists on site, with information provided concerning the Sweet Track's discovery and modern reconstruction (Bruning 2006). The reconstruction is a major draw for visitors (see Image 3.3) and helps illustrate the Avalon Marshes Centre's storyline that the Somerset wetlands have been important to human development, not just for livelihoods but also for community building.

Image 3.3 Sweet Track reconstruction, Shapwick Heath, Somerset. (Credit: Avalon Marshes Centre)

This is true too of Alkborough Flats (Fig. 3.2), where the local council uses the antiquity of the site to encourage visitors to this relatively out-of-reach part of England. Key to this narrative presentation is the palimpsest nature of the space; used in different ways over millennia by humans. The Neolithic, Iron Age and Roman era are all represented on the information board, highlighting the utility of the site in different ways. As ever, the depictions on both these information boards create a heteronormative vision and version of the past, which is presented as factually accurate. In truth, much post-processual archaeological interpretations suggest that women had equal status in prehistory with men. This alerts us to always question normative representations of the past (Gilchrist 2012).

As our arguments indicate, family play activities form a central element of these spaces. In Bedfordshire they are a central rationale for creating and maintaining these sites. They are presented as quasi-public goods—spaces for wildlife and for human recreation. PCP is a council-funded site, and as a result one element of their purview is to provide recreational amenities for local residents (while also, of course, defining a physical and social barrier between the residential and rural wetland environments). The online pages of PCP sit within the council's 'culture and leisure' division. Parking on site is free, toilets are open, paths are well maintained, and there is an independent café operating on site. The Council rangers are also encouraged to provide outreach activities with different publics. Guided walks,

Fig. 3.2 Alkborough Flats Heritage trail: visitnorthlincolnshire.com

park runs and half marathons are all standard provisions. The add-ons depend on ranger skill sets. In the past they have included guided bat walks, pond dipping, bug exploring and mushroom foraging. As discussed in Chap. 2, austerity politics has, even during the time of the WetlandLIFE project, reduced ranger numbers by two-thirds. As a result these activities which serve to really embed experiential learning and ludic play within wetland environments have ceased to be offered on this site. For MCP, as a charitable trust, the operating conditions are different. Their rationale is to develop a community wetland site as part of a Community Forest endeavour, within a wider landscape mosaic (see Chap. 2). MCP wish to encourage a diverse visiting public but also need income-generating activities; so alongside the visitor's centre acting as a conference venue, it is also hired out for weddings and community events. Combining play and income are activities such as bike hire, pond dipping and foraging—with all proceeds being funnelled back into the not-for-profit activities on the site.

The two other wetlands are markedly different. In Somerset family play activities are confined to the trackways, with cycling, walking and scooting permitted. These are shaped around the created contours of Westhay Moor and Shapwick Heath and feel markedly less sculpted and 'designed in' than in Bedfordshire. The message here is that human activity fits within the renaturalised sites to develop a space that is as rewilded as possible within a managed environment (Hall 2016). There are no play parks, cafés or toilet facilities, so the time spent here, especially for little ones needing toilet access, is inevitably shortened. The same is true of Alkborough where amenities are limited. Families can make use of the private tea rooms close by for toilet breaks and refreshments, but the lack of any visitor's centre creates a space in which few families were observed to dog walk, scoot or stroll by the expanding reed beds during the empirical fieldwork. Alkborough feels very much like a specialist's destination—ideal for those looking for particular bird species or antiquarians looking to visit Julian's Bower. The management of the site is overseen by a multi-partner committee comprised of the North Lincolnshire Council, the Environment Agency and Natural England, amongst others. Their rationale, as we've seen in Chap. 2, is to coordinate the managed realignment scheme alongside developing strong community relations. Family play is not at the centre of this organising framework; indeed, the layout and management of the site does not provide for family play spaces.

Playing Up: Specialist Recreation

Consistent with Blackshaw's (2010) arguments, much of the serious business of play in these wetland sites is focused on the specialist—the birder, the wildlife photographer, the walker, the artist, the naturalist, the angler, the cyclist, the sailor. These activities are not unique to wetlands, but they are ideal for these waterscapes, with the spaces offering sites in which these practices, and their associated identities, can be developed and honed. Exogenous skills and equipment are brought to the site, often expensive, technical equipment which demonstrate the seriousness and prowess of the individuals, and which also require that these spaces are perceived as safe.

Many of the interviewees discussed the diverse attributes of the sites as enabling them to perform and enjoy their recreational activities in certain ways. For one female birder on the Shapwick Heath site, the solitude of the site was an enabler, not a barrier, to solo excursions:

> *I prefer to come early in the mornings because, well, you sort of think the people who might break into cars aren't there early.* (Female birder, Somerset, 60s)

Another female recreationist, an artist, discussed how the quietude of early morning was necessary for her practice. She felt connected to the trees, green paths and wildlife at these moments, alone with her sketch pad, often audio recording bird calls and reeds waving in the breeze. This is true too of a spiritual practitioner, who stated that the walk to and from and around the site gave her just enough time to recharge. She craved the anonymity the space could give her outside of a role where public and private collapse:

> *I think there is a sort of sense of ownership and affection for the place, even though it's not sort of pretty in the conventional sense. I like it in the depths of winter actually because it's just so austere.* (Female spiritual practitioner, Alkborough Flats, 40s)

For many these sites provide solitude. Bird hide etiquette requires limited conversation, with collective silence the overriding expectation so that a churchlike ambience is created in these spaces for quiet contemplation. Understanding and observing this etiquette is very much part of the practice for such visitors. Outside, birders and photographers often baffle their interaction with others through the equipment they use, adjusting lenses and binoculars to deter conversations. Many interviewees discussed how their love of nature began in their childhood, and moreover that their ability to cope with the demands of everyday life was made possible through the restorative possibilities of their wetland activities. Especially in the Alkborough site, these ludic activities are precious—the time spent driving to and from the site being part of the recreational practice itself—as explored by Marion Clawson (1959) in one of the first studies of leisure behaviours.

Not all recreational practices that could be undertaken in these spaces are permitted. The most permissive site is the PCP, though it has by-laws which prevent mass gatherings, barbeques, loud music and other social activities which take place in other public spaces. Like many wetland spaces which engage diverse publics, the nature reserve status limits forms of collective engagement in these places (Gearey et al. 2019). However, some interviewees felt that some wetland managers were not willing to engage with community outreach activities. In the Somerset site, one waterways organisation felt that there was a policy of passive non-engagement, demonstrated through both a lack of outreach from the SWT itself and non-responsivity to enquiries. They felt that the position was wildlife first, humans second—and so a 'fortress conservation' mentality prevented considerations of enabling more diverse ludic play on the site. This particular organisation felt the space was ideal for differently abled recreationists. They wanted to develop an approach:

> ...which is about encouraging people with disabilities. So things like wetlands actually could be a really good environment if somebody's got considerable visual impairment... hearing birdsong could be quite a benefit.

THE LUDIC IMAGINARY: SENSE OF SELF, ADULT AS CHILD AND TEENAGE KICKS

Aside from the physicality of being in wetlands, such spaces also play an important role in the imaginary as places in which a remembered or fictionalised version of self is replayed. Caillois (1961) in his study of agency in play differentiated *paida*, anarchic, unformed expression as play, as opposed to structured, goal-orientated *ludus*. For some of our WetlandLIFE interviewees, recalling past times and past activities in wetland spaces remains a vibrant part of their current self-identity—tapping into a more free form expression of self, where desire, morality and shame are a little less prevalent, a critical eye cast on society's expectations. Rather than focus on current physical ailments, or difficulty in accessing these spaces, the interview participants highlighted how their current sense of self and delight in nature was forged through former interactions on these sites, often connecting with positive associations in childhood:

> *I was at the Forest Centre yesterday and they've got a forest school out in the wood, it's a din, a din, but hey, they were happy and they were all running about and all engaged. They'd been pond dipping and they were so happy, they weren't grumpy or grisly, they were just happy, running about and it's the way in, if you start them early, give them some knowledge and that's what got me in the first place, my dad and his endless stories about creatures.* (Volunteer, female, Bedfordshire, 70s)

Many were very grateful for the freedom that the wetlands provided as an 'othering' space. Adopting the role of the volunteer, the fund-raiser, the walking group member and the community organiser, amongst others, all enabled another side of their personalities—the playful, considerate, not-fully-formed part of themselves—to run free. Casting back to earlier times, some of the more senior members of the interview group recalled friendships formed and physical challenges undertaken in these spaces: tree planting, the clearing of invasive species, leading bug hunts and other guided events:

> *I can remember coming swimming down here in the early seventies when it was the pits, as a young teenager, probably with Ian, in Boys Brigade we had a tour of the power station which was where the River Field Estate now is, and on the grass down there there is still marked out circles of one of the, chimneys.* (Volunteer, male, Bedford focus group, 50s)

This volunteer, who swam in the gravel pits before their transformation into the water reserves for Priory Country Park (PCP) is now a grandfather who takes his grandkids to the wetlands for otter and kingfisher spotting and to windsurf and paddle board at the marina. This link between playing as a kid and taking his family to play on the site highlights the intergenerational linkages of a sense of place.

This 'othering' space is also important for another, maligned aspect of childhood: adolescence. It seems as if in early childhood and adulthood wetlands enable prescriptive, 'allowable' ludic practices to play out. For young children, pre-teens, the ludic is a Blakean feast of innocence (1998)—pond dipping, bug hunting, nature explorations. Adulthood is riven with the seriousness of the expert—nascent birders, walkers, naturalists, bat fiends. In between lies the delinquent. First the teenager, a slave to hormones and neuro-plasticity which renders them part child, part proto adult. Young teenagers in nature can be seen as a threat to the neoliberal order as their consumption activities drop away as they spend less time with adults and family members; they are less likely to be purchasing, they have more thinking time, they are creating free forms of social solidarity beyond adult control. In wetland spaces they can assert their own agency, even if elements are destructive and wanton: alfresco sex, raves, fly-tipping, bird poaching, drug dealing, rough camping—forms of communing often shut out from many urban spaces (see Ravenscroft and Gilchrist 2009). All these have been documented in the study sites. Whilst they are nothing new to wetlands, they rarely are mentioned in any forms of literature or reviews. It seems we do not want to mix our bucolic with the lived reality of complex lives, where less managed spaces invite the possibility, or the attainability, of othering forms of social behaviours which are messy, complex and, most importantly, reflect back to us the contested nature of our relationship with each other:

> *I just, I just camp, just wherever, let's see. We started off Alkborough, walked down, past Grimsby to, towards Tetney Lock, and by Nook where there's a big bombing range and it's all country so we watch, I was watching the Osprey helicopters do practice fire, you know, on the sands there whilst the Pride of Rotterdam ferry was sailing out of the mouth of the Humber.* (Male psychogeographer, Alkborough Flats, 30s)

Before exploring the adult ludic play which is linked to differing forms of the wetland imaginary, it is important to dwell on teenage kicks. Increasingly in our neoliberalised world, we don't quite know what to do with teenagers. Generation Z, the millennials on the cusp of adulthood, is a market which has been shaped since birth (see Williams and Page 2011). For the purposes of wetlands, we need to understand how youth culture, or youth branding, impacts the use and value of these spaces.

Buckingham and Willett (2013) have argued how far our children have become digitised—hybrid cyborgs attached to phones, tablets, MP3 players and other devices continually streaming information in and out. Kids are data reservoirs and channels—the impossibility of imagining teenage play outside of domains of content provision/marketisation means that wetland managers are struggling with how to work with this cohort in a space where things are slow moving and not a great deal happens. Although much attributed to him, but never claimed, Fredric Jameson has suggested, with his hat tipped towards both J.G. Ballard and H. Bruce Franklin, that it is easier to imagine the end of the world than it is to imagine the end of capitalism (Teeuwen 2009).

Added to this is a closure of places and ways of being together for adolescents. Digitised worlds enable them to share and communicate online; places of physical indoor interaction remain shopping centres, fast food outlets, cinemas, arcades and, influenced by online pornography and Instagram feeds, the gym. Yet in post-austerity England since the 1980s, there has been a steady decline in youth spaces—youth clubs, social clubs and theatre and arts clubs. The era of hanging about in sight of socially minded adults has dwindled; another facet of Austerity politics, the closure of comunity spaces. Instead, ever ingenious and creative, young people are making new attachments in outdoor spaces. Raves, outdoor parties, congregating in parks and beaches, Facebook meet-ups in houses and renting of Airbnbs for mass events become the new norm. This is true for wetlands too.

Increasingly, particularly in the urban and semi-peripheral wetlands, these less managed spaces become retreats for gatherings. In the urban wetlands, the bird hides have been the centre for group parties. Most activities are—benign; beer cans and cigarette butts are left behind. At worst hides are graffitied, burnt, left littered with nitrous oxide canisters and balloons, used condoms and rejected clothing. Site managers are often suitably enraged, but still do take great steps to manage the cause of the problems through trying to work with local youth communities so that they learn to respect the space. In many ways—and consistent with practices at other countryside sites (see Ravenscroft and Gilchrist 2009)—these managers don't mind what happens in these spaces out of hours, as long as a 'leave no trace' policy is followed.

The same is true for outdoor parties and the use of the space for get togethers, camp fires and wild camping, skinny dipping, drug taking. For one interviewee, the teenage drug taking that formed his adolescence in Lincolnshire has informed his ongoing practice as an artist. Rather than psychedelic or intuitive ramblings around space, he sees his forays into wetlands as both delving into the immediate landscape and into deep time. Initiated through psychoactive drug taking years ago, he feels this forged a deep connectivity with landscape, enabling him to view spaces often seen as barren and wasted—the quintessential proto-modernist view of wetlands—instead as dynamic environments which are captivating, intimate and endlessly fascinating. For him, teen years are about exploration of the self, with wetlands as the ideal less managed space in which to do this. Accessible for city kids and rural teens, providing a form of equitable bacchanalia, wetlands enable a much needed outlet for crowded out and data bombarded youths. It is a creative, meaningful play, a voyage of inner exploration that can ultimately lead to forms of living with the Earth which is more patient, less demanding. The interviewee's current artistic practice attests to these early ludic forays into his psyche, leading to a form of lived, authentic spiritual connectivity which has humour, pep, rigour and boundless drive.

The wetlands here involve truly free play, which in many ways is boundary testing. As we explore more coherently in Chap. 5, wetlands as liminal spaces, prevalent throughout diverse human cultures, remains true today. Arcane practices of bog gifts, both as artefacts and as human sacrifice, preserved in the oxygen starved peatlands of Northern Europe, remind us that these spaces were privy to a whole range of othering practices that sit outside of our view of what can be designated 'normal'. We can describe these practices as delinquent, inasmuch as their logic is framed by different value systems and practices. The WetlandLIFE project has documented a whole range of othering activities in the case study sites—rough sleeping communities; cruising, dogging, flashing; drug dealing; industrial scale fly-tipping and domestic rubbish dumping; irresponsible dog walking businesses leaving dog excrement and harassed bystanders in their wake; waterfowl and fish poaching; bird egg stealing; mass litter left by cyclists and runners; graffiti and arson.

Some have more alternative ludic interpretations than others. The alfresco sex, graffiti and arson fit within this category of displaying attitudes and behaviours which enact self-determination and self-expression, even if these are socially compromising. The other forms of ludic use of wetland space connect with wider contemporary social and cultural issues. Rough sleeping makes sense, particularly in the urban wetlands, as these secluded wooded spaces offer refuge and a degree of safety and privacy not easily found in built environments. These are pragmatic responses to a dreadful lack of social will, and political and economic withdrawal, to tackle complex issues regarding mental health, employment and housing amongst other nested contributory factors. For the people living in these sites, they may regain even a small element of agency in these troubled times.

The fly-tipping, domestic waste disposal and petty user littering—often of gel packs, protein bar wrappers and energy drink cans—have the same environmental impact but different drivers. Looking to the sports users as unexpected depositers of detritus, one can only speculate that there is a presumption that the roadside verges are regularly cleaned by the Highways Agency or local councils. Instead it is groups of volunteers, from the local community, working with the sites, or from the Guides or Scouting Association, who undertake the litter picking.

Intertextuality—Wetlands as Online Spaces, Spaces of Spectacle

I really appreciate that the Levels are full of beauty as an artist, you know, for inspiration I can go down there, I can meet huge groups of people, really fascinating, you know, you can learn so much, starling sell all day long, if you want to do a picture, do a starling, it's great. (Artist, Somerset Levels, female, 50s)

Sex sells, but starlings sell even better. Starling murmurations are the unofficial emblem of English wetlands—a crepuscular aurora borealis of animal bodies in co-responsive flight patterns. The murmurations of these over-wintering birds begin in mid-autumn and continue until spring, as starlings make their way from their Eastern European summer abode in Russia and Ukraine. Their mesmerising, shape-shifting roosting in groups of ten thousand or more attract crowds of terrestrial onlookers. Once the preserve of the birding community, these murmurations now are a draw for tourists, who travel to the sites specifically to witness these avian gymnastics. They are enabled through wetland websites which advertise these natural events as part of the fabric of the autumn and winter schedule of the site, along with phones numbers to access prerecorded messages which provide GPS data for likely roosting sites each day.

These murmurations are digitised spectacle. Vblogs are uploaded, tweets dispatched, Snapchat stories shaped and are the making of Instagram heaven. The affordance of the starling's natural response to eluding quarry is anthropomorphised into repositioning their actions as ludic—playful, joyful, a wonder of nature. To watch their swirling flight patterns is to witness a display of natural wonder. Yet the display is precautionary, intrinsically one of survival, as the outer lines of the murmuration warp and weave as the birds strive to reposition themselves more centrally in the hub for their own self-protection. Ultimately we are witnessing survivalist strategies which we recast as ludic. Animals in nature do not waste calories without good cause.

Repurposing survival activities as playful enables diverse forms of income generation in English wetlands. The wetland sites utilise the car parking fees of murmuration visitors as primary funding, with the uploaded texts, tweets, blogs and instafeeds as social promotional tools for future visitors, both within the UK and overseas. The murmuration industry also supports local incomes through tourist spending on accommodation, food and ancillary purchases of professional photos and artwork in nearby artist spaces. Birders attending murmurations with starling tourists describe ways in which these incomers chat ceaselessly, with whoops and cheers as the birds whirl around. A very visceral part of the experience is lost—the sound of thousands of wings moving in unison close by, changing eddying air currents and creating hypnotic aural patterns, as the wing beats echo off trees and water. Seasoned ornithologists document where the birds roost at night—and instead attend the reverse process, the dawn murmuration as the birds lift as one in the morning light to find that day's feeding ground together. This time without the selfie sticks and expensive cameras.

Murmurations create online and semi-fictionalised wetlands. Judicious editing makes every wetland attendance appear potentially transcendant and profitable in the pursuit of process capturing the spectacular. Filters and photoshopping edit out the messy elements of these events. In the online world, starling roosting is mesmerising and enchanting. The real-life travails of parking in muddy, poorly paved car parks, weeing in bushes, standing often for hours in the cold and murk, and waiting for capricious bird arrivals that ultimately roost elsewhere, are all smoothed out and obfuscated. The natural element of these processes, and their inherent unpredictability, is at a remove as the digital world enables us to create a wetland of our imagination, as a safe harbour for these tiny, amazing creatures who fly 6000 miles in the course of a year. The wetlands become a fabricated wildscape of sorts, where environmentalists can present these events as impossible without particular types of land and water management regime. Their aim is to ensure that wetlands remain consistently the same year after year—even though without human intervention, these starlings would adapt to change and relocate according to natural variation. But those who stand to profit, in the widest sense of the term, from the murmurations—the wetland managers, the local artists and the local economy—need certainty against the vagaries of real-world discontinuity. The ludic is a very managed, artificial goal which may, in terms of intragenerational aspects, work to the detriment of the starling species.

Birds are as epigenetic as humans; they learn, adapt and forget. There is much to speculate as to whether creating highly managed wetland spaces, such as the three case studies in our book, will ultimately create biodiversity enclaves. If we are mistaken in thinking that we protect our more-than-human brethren through forms of 'fortress conservation', that is, that we can use wetlands as safe spaces against the general decline in overall biodiversity, these isolated havens will have the opposite effect of reinforcing these as sites of ludic spectacles of nature, divorcing us further from our own elemental immersion into nature. Creating these biodiversity barbicans salves our conscience, even as global ecosystems are destroyed through processes linked to climate change, neoliberal consumerism, population growth and anthropocentrically driven environmental degradation.

Spatiality and context are key considerations as we contemplate the starling. In the USA they are designated a pest; all thanks to the romantic actions of Eugene Schieffelin who released the first European starlings into North America (Raber 2015). His mission? To provide the New World with every bird listed within Shakespeare's canon. The ludic becomes lunacy. These non-native starlings have since overwhelmed domestic species as the dominant predator, decimated the Wheat Belt of the USA and overrun urban spaces with their swarm mentality, creating havoc in domestic gardens, their acidic droppings toxic to humans in dense concentrations. We should marvel at the murmuration, but remind ourselves of the foxes, owls, badgers, rats that rely on these starlings within the food chain. They are watching the murmurations too, with quite different eyes; and we should be thankful to them.

Playing for Keeps; Wetlands as Inclusive Ludic Spaces

By way of conclusion, we reflect on wetlands as recreational spaces and who should benefit from them. The answer is complex; benefits suggest that wetlands themselves have no agency—that they exist for human development only. Of course, they exist for no one; they are a series of complex, integrated ecosystems containing many different species that are, to greater or lesser extent, managed for society, for the economy, for the environment. They can also clearly be understood from an ecosystem services perspective as providing benefits to humans through play experiences and learning. Exploring the ludic quality of our interactions with wetlands, however, takes us beyond the focus of ecosystem services and provides us with the ability to both consider their current use by particular socio-cultural groups and to use this reflection to explore ways of making them more inclusive. The wetland managers were acutely conscious of the tensions that can arise between different users seeking ludic and wellbeing experiences. The emphasis on families and particular types of play sets norms about acceptable behaviour and creates boundaries for user behaviour that may also discourage others. Activities involving drugs, sex and graffiti perceived by some users as deviant are also accommodated to some degree in the dusk or dark as long as these remain spatially confined to certain locations and that nature is not harmed through littering or vandalism. The significant conflicts that have arisen on other inland waters (Church and Ravenscroft 2011) were not evident in these different forms of wetland either because some users are not permitted or in the case of canoeists and anglers are spatially managed into designated zones and times.

Conflicts, inclusion and exclusion can be seen as highly human concerns, but the ludic activities we find in wetlands highlight the agentic aspects of the more-than-human. Fauna and flora, especially birds and murmurations, interact with humans to shape the ludic experience. As a result the behaviour of certain species, through attracting large numbers of visitors, will challenge the biodiversity concerns of wetlands, resulting in managers taking actions to protect the environment that may encourage some social groups to use wetlands and deter others. This sculpting of wetland space for particular social groups is not a contemporary phenomenon, as we explore in the next chapter, interrogating the place of literature in making, and responding to, wetland environments. And so as we move from play in this chapter, we move into a play within the next.

References

Blake, W. (1998). *Songs of Innocence and of Experience* (Vol. 5). Princeton: Princeton University Press.
Blackshaw, T. (2010). *Leisure*. Abingdon: Routledge.
Bruning, R. (2006). *Wet & Wonderful. The Heritage of the Avalon Marshes*. UK: Somerset Heritage Services.
Buckingham, D., & Willett, R. (2013). *Digital generations: Children, young people, and the new media*. London: Routledge.
Caillois, R. (1961). *Man, Play and Games. Translated by Meyer Barash*. New York: Free Press.

Carothers, P., Vaske, J. J., & Donnelly, M. P. (2001). Social values versus interpersonal conflict among hikers and mountain bikers. *Leisure Sciences., 23*, 47–61.

Chubb, M., & Chubb, H. R. (1981). *One third of our time? An introduction to recreation behavior and resources*. New York: John Wiley & Sons Inc..

Church, A., & Ravenscroft, N. (2011). Politics, research and the natural environment: the lifeworlds of water-based sport and recreation in Wales. *Leisure Studies., 30*(4), 387–405.

Church, A., & Coles, T. (2006). Tourism, politics and the forgotten entanglements of power. In A. Church & T. Coles (Eds.), *Tourism, Power and Space* (pp. 1–42). London: Routledge.

Church, A., Ravenscroft, N., & Gilchrist, P. (2013). Property ownership, resource use and the gift of nature. *Environment and Planning D., 31*(3), 451–466.

Clawson, M. (1959). The crisis in outdoor recreation. *American Forests, 65*(22), 40–41.

DEFRA. (2018) Landscapes review: National Parks and AONBs Led by Julian Glover. Published online 27th May 2018. https://www.gov.uk/government/publications/designated-landscapes-national-parks-and-aonbs-2018-review. Accessed on 3rd October 2019.

De Grazia, S. (1964). *Of time, work and leisure*. New York: Twentieth Century Fund/Anchor Books..

Díaz, S., et al. (2018). Assessing nature's contributions to people. *Science, 359*, 270–272.

Fish, R., Church, A., & Winter, M. (2016). Conceptualising cultural ecosystem services: a novel framework for research and critical engagement. *Ecosystem Services., 21*(B), 208–217.

Gearey, M., Church, A., & Ravenscroft, N. (2019). From the hydrosocial to the hydrocitizen: Water, place and subjectivity within emergent urban wetlands. *Environment and Planning E: Nature and Space, 2*(2), 409–428.

Gershuny, J. (2000). *Changing times: work and leisure in post-industrial society*. Oxford: Oxford University Press.

Giblett, R. (2009). *Landscapes of culture and nature*. London: Springer.

Gilchrist, R. (2012). *Gender and archaeology: contesting the past*. London: Routledge.

Hall, M. (2016). Extracting culture or injecting nature? Rewilding in transatlantic perspective. In M. Drenthen & J. Keulartz (Eds.), *Old world and new world perspectives in environmental philosophy* (pp. 17–35). Berlin: Springer International.

Hemingway, J. L. (1988). Leisure and civility: Reflections on a Greek ideal. *Leisure Sciences, 10*(3), 179–191.

Huizinga, J. (1950). *Homo Ludens: A study of the play element in culture*. Boston: Beacon.

IPBES. (2019). *Global assessment report on biodiversity and ecosystem services of the Intergovernmental Science-Policy Platform on Biodiversity and Ecosystem Services*. Bonn, Germany: IPBES Secretariat.

Jacob, G. R., & Schreyer, R. (1980). Conflict in outdoor recreation: a theoretical perspective. *Journal of Leisure Research., 12*, 368–380.
King, K., & Church, A. (2015). Questioning policy, youth participation and lifestyle sports. *Leisure Studies., 34*, 282–302.
Mair, H. (2002). Civil leisure? Exploring the relationship between leisure, activism and social change. *Leisure/Loisir, 27*(3–4), 213–237.
Oncescu, J., and Neufeld, C. (2019) Low-income families and the positive outcomes associated with participation in a community-based leisure education program, Annals of Leisure Research 1–18.
Opaschowski, H. (1998). The future of leisure, culture and tourism – the challenge for politics and society. In W. Nahrstedt & T. Pancic Tombol (Eds.), *Leisure, culture and tourism in Europe: the challenge for reconstruction and modernisation in communities* (pp. 15–26). Institut fur Freizeitwissenschaft und Kulturarbeit: Bielefeld.
Ravenscroft, N., & Taylor, B. (2009). Public engagement in new productivism. In M. Winter & M. Lobley (Eds.), *What is land for? The food, fuel and climate change debate* (pp. 213–232). London: Earthscan.
Roberts, K. (2011). Leisure: the importance of being inconsequential. *Leisure Studies, 30*(1), 5–20.
Rodriguez, H. (2006). The playful and the serious: An approximation to Huizinga's Homo Ludens. Game. *Studies, 6*(1), 1604–7982.
Rojek, C. (2001). Leisure and life politics. *Leisure Sciences, 23*, 115–125.
Rojek, C. (2010a). Leisure and emotional intelligence. *World Leisure Journal, 52*(4), 240–252.
Rojek, C. (2010b). *The labour of leisure: the culture of free time.* London: Sage.
Raber, K. (2015). Shakespeare and Animal Studies. *Literature Compass, 12*(6), 286–298.
Ravenscroft, N., Uzzell, D., & Leach, R. (2002). Danger ahead? The impact of fear of crime on people's recreational use of non-motorised shared-use routes. *Environment and Planning C: Government and Policy., 20*(5), 741–756.
Ravenscroft, N., & Gilchrist, P. (2009). Spaces of transgression: governance, discipline and reworking the carnivalesque. *Leisure Studies, 28*(1), 35–49.
Sumner, J. and Mair, H., (2017). Sustainable leisure: building the civil commons. *Leisure/Loisir*, 41(3), 281–295.
Teeuwen, R. (2009). Ecocriticism, humanism, eschatological jouissance: JG Ballard and the ends of the world. *Tamkang Review, 39*(2), 39–57.
Veal, A.J., (2012). The leisure society II: The era of critique, 1980–2011. *World Leisure Journal, 54*(2), 99–140.
Williams, K. C., & Page, R. A. (2011). Marketing to the generations. *Journal of Behavioral Studies in Business, 3*(1), 37–53.
Woodyer, T. (2012). Ludic geographies: not merely child's play. *Geography Compass, 6*, 313–326.

CHAPTER 4

Wetlands as Literary Spaces: Off Kilter, Off Grid, Off the Wall

Abstract The chapter explores the dialectical relationship between literature and wetlands, with a very particular English bent. The renaissance in new nature writing, and its current vogue as nature pilgrimage mixed with personal, often pain, memoir, has led to an abundance of accessible, and beautifully written, paeans to different English landscapes. But there is a 'New Folk Horror' revival too, also known as the English Eerie, providing an alterantive 'off kilter' perspective on our relationship with natural spaces. This chapter will explore this fascination with the dark side of the English pastoral in all its artistic forms such as filmmaking, music, art and storytelling. Changes in modern publishing have also enabled literature to go 'off grid' as well as 'off the wall'. The chapter ends with a reflection on the key influences to our modern cultural representations of wetlands.

Keywords Dialectical • New nature writing • New Folk Horror • English Eerie • Cultural representations

> *You nimble lightnings, dart your blinding flames*
> *Into her scornful eyes! Infect her beauty,*
> *You fen-suck'd fogs, drawn by the pow'rful sun,*
> *To fall and blast her pride!*
> King Lear, Act 2, scene 2

We move from the previous chapter into this with our own ludic gambol. From thinking about real-life play to considerations of the ways in which art reflects life or, at very least, presents to us slightly altered versions of reality. Does life inform art—or the other way around? We start by using lines from a play, Shakespeare's *King Lear*, to see the way in which culture, the arts, can influence our perspectives on landscape. Even here, nearly five hundred years ago, the marshes are the spaces of earth magic, places where natural forces can be called upon to wreak havoc and unsettle. Wetlands have long been held in regard as eldritch landscapes of *The Weird and the Eerie* (Fisher 2016).

Who, depending on your age, does not recall the shiver down the spine on first encountering Dickens' Magwitch (1993)? Or watching through fretted fingers the original *Hound of the Baskervilles* with Basil Rathbone, possibly on rainy Saturday afternoon TV? Pulp fiction and serialised stories in penny dreadfuls drew the reader in with tales of the macabre and the fearful. Accompanying pictures of foggy moors and benighted travellers recoiling from monstrous beasts (see Image 4.1) and depicted in garish colours, ensured reader's appetites were momentarily sated, until the next weekly or monthly instalment. If we reflect on the image below a little more, its resonance with modern cultural imagery is revealed. The book's central characters, Watson and Holmes, are rough renditions of the film actors who brought the detective series to a new audience in the 1930s, complete with their onscreen deerstalker hat and tweed suits. Chiaroscuro techniques are deployed, where an unholy light appears to emanate from the black shuck, against the darkness of the elemental moor. The hound's jaws are dripping blood from a previous unseen victim, like some Gothic vampire horror. The curls of fog shrouding the protagonists directly recall folk tales of will-o'-the-wisp, luring innocents to their doom. So, combined, we have modernist film references, Italian renaissance painting effects, Victorian hysteria and folk tales—all bundled together in one cheap book jacket cover.

Maybe your literary experience of wetlands as spaces of horror is more recent, leading you to consider whether the Essex Serpent (Perry 2017) was real or the hallucination of Victorian hysteria; whether Daisy Johnson's characters (2016) in *Fen* are those drawn to these marginal spaces—or are the product of these landscapes. Your reference points for beasts, wretched mortals or the ethereal 'other' that dwell within wetlands may be more classically framed: Beowulf's Grendel (Heaney 2009), Bunyan's bedraggled pilgrim (1967), Poe's Sphinx (1914),

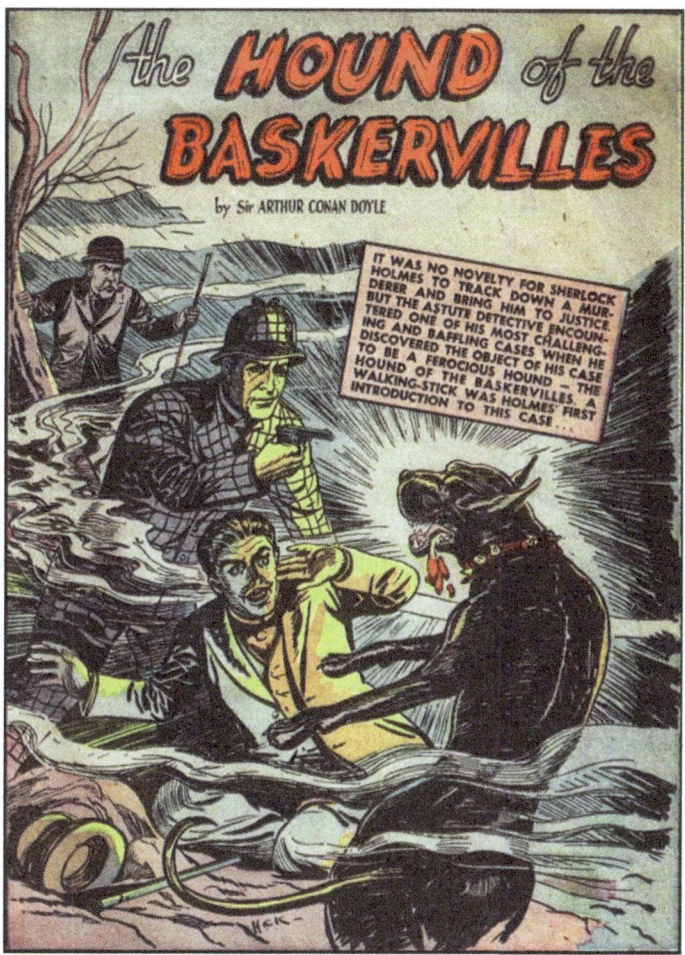

Image 4.1 A classic front cover jacket for crime novel *The Hound of the Baskervilles by Sir Arthur Conan Doyle*

Tolkien's Gollum (2012), Jonson's Fitzdotrrel (1994), Shakespeare's Caliban (1964). Combined they are a hybrid tangle of wayfaring travellers and clueless outsiders, lost and disorientated in these spaces alongside the manifestations of distorted man-beasts and bog sprites rendered inhuman by the fens, swamps and moors they inhabit.

Your tastes may be more towards the Romantic; Wordsworth, Keats, Bronte and Blake (see below) all found the sublime within their English wetlands, so close to home. It may be that your reference points sit outside of the UK—the work of Walt Whitman, Emily Dickinson, Ernest Hemingway and Willa Cather are all shaped by the swamplands of their habitus, with an earthy, sensual connection to the hollows and mires in their backyards as the setting for self-reflective essays and poems, and for stories of homely folk and lives in microcosm. William Blake's ecstatic visions of heaven and hell operating side by side in the landscapes of his then pastoral North Lambeth home in Greater London also reflected his feelings towards secular society and his sympathies towards revolutionary ideals. Consider his poem 'The Garden of Love' below, which utilises the motif of the water meadow, at first comforting, and then a wasteland as nature is transformed by human agency. Blake was able to see the world transforming around him. His work anticipated the rupture that rapid industrialisation and the associated orthodoxy of organised religion was to bring to landscapes, shared lives and to our spiritual selves. The way we view nature is shaped by what we think it is 'for'; and these perspectives are shaped by the cultures we are embedded within. This chapter will ask you to consider how far our perceptions of wetlands are shaped by literature, and art more generally, or whether the wetlands themselves, and the folk tales, oral traditions, political associations and local hearsay associated with them, truly shape the stories that emerge in our culture (Garden of Love, William Blake 1994).

> *I laid me down upon a bank*
> *Where love lay sleeping*
> *I heard among the rushes dank*
> *Weeping Weeping*
> *Then I went to the heath & the wild,*
> *To the thistles & thorns of the waste*
> *And they told me how they were beguiled*
> *Driven out & compelled to be chaste*
> *I went to the Garden of Love,*
> *And saw what I never had seen:*
> *A Chapel was built in the midst,*
> *Where I used to play on the green.*

As well as a meander through a treasure trove of literary gems, and a reintroduction to some hard to forget fictional and folklore characters, the chapter will explore how far the positioning of wetlands as dank, dangerous spaces intentionally supported the continuation of a wide number of delinquent and illicit practices on these sites. From the romanticised smuggling, poaching and distilling of past times to more modern 'other uses' such as rough sleeping, teenage partying, magic mushroom foraging, public sex and drug dealing, the 'wilderness' provided by wetlands still retains a modern darkness which changes in literary appetites still find hard to shake off.

Literature has still kept its grip on the darker side of wetlands, but the changes in modern publishing are reflected in wetland literature too. The renaissance in modern nature writing and its current vogue as nature pilgrimage mixed with personal, often pain, memoir has led to an abundance of accessible, and beautifully written, paeans to different English landscapes. Wetlands have had their admirers (Deakin 2011; MacFarlane 2012; Laing 2011), but none have used wetlands solely as their muse.

Alongside nature writing, and sometimes with the same protagonists—such as McFarlane—are those who celebrate the arcane, Gothic and occult aspects of the English landscape. Drawing on the classic 1970s and 1980s literature of Alan Garner and Susan Hill, themselves drawing on the work of Algernon Blackwood, M.R. James, James Buchanan and others, there is a 'New Folk Horror' revival: also known as the English Eerie, the English Gothic and Pastoral Noir. We will explore further this fascination with the dark side of the English pastoral in all its artistic forms such as filmmaking, music, art and storytelling.

Changes in modern publishing too have enabled literature to go 'off grid' as well as 'off the wall'. E-publishing, blog and vlog sites, Twitter hashtags to share memes and videos, crowd-funding links and many other digital and community forums have enabled aspiring authors to self-publish. For wetland literature this has spawned a series of sub-genres—historical fiction (Muir's 2015 and onwards *The Ghosts of Culloden Moor* series), children's adventure stories (Pullman's *His Dark Materials* trilogy, 1997, and later *Book of Dust* prequel, 2017) and even a raft of libidinous zombie and vampire tomes (Gill 2013). Both the English Eerie and the self-published novels act as an engaging counterpoint to the benign, wholesome nature of nature writing.

These serve to tell us about the fractured identities of wetlands in modern contemporary English culture. Part wastelands, part beloved wilderness, they remind us that beauty is in the eye of the beholder—and that cultural representations take a long time to become reshaped. Using the case study sites for reflection, the chapter reviews the diverse socially generated prompts and drivers that local people cite as impacting on the use and value of their local wetlands from a cultural context.

The chapter ends with a reflection on the key influences on our modern cultural representations of wetlands. These include drivers that are macro in scale, and are hence often occluded from everyday purview, such as policy, legal or institutional decision making around English wetlands, and those which are grassroots in origin and more immediate and impactful, such as the heuristics, the stories, the people, involved in day to day wetland management and practices.

Classical Fictional Literature

Whenever we talked with people about the WetlandLIFE project, particularly with regard to the wetlands of our imagination, many shared similar reference points. Aside from Dickens and Conan Doyle, most people cited *Wuthering Heights* and *Jamaica Inn* as their pivotal English wetland texts. Both *Wuthering Heights* (2009) and *Jamaica Inn* (1936) are linked in the readers' mind with wetland landscapes of wild moors. Cathy and Mary's stories could only have emerged from these wild, uncivilised spaces. The former is closely linked with Haworth Moor in West Yorkshire, the second with Bodmin Moor in Cornwall, and thanks to the success of the books, and their subsequent black and white films made in the heyday of British Cinema, they are also important for current local heritage industries. Bleakness is big business.

Emily Bronte's paean to romantic love was, of course, also immortalised by the songwriter and artist, Kate Bush. *Wuthering Heights* calls to a fascination with wild places and unhinged romance. As we will see throughout this chapter, there is a particular fascination with the darker side of our relationship to landscape which has forged types of British artistic life. In writing, music, art and other forms of self-expression, wetlands are fundamentally part of a cultural legacy, which has shaped how we see these spaces in our mind's eye, even if they are spaces we have never visited. Wild and wet moors then become the spaces for dark and desperate acts, sadly not always fictional.

How far do these stories utilise established fables, rework traditional tales or provide us with new contemporary mythologies? The contrary roles within folklore in replicating social norms, subverting cultural expectations and encouraging a mindfulness in nature are well established. Stories, songs and other oral traditions enable the transition of information without the need for written language or documented processes. Robert Macfarlane's *Underland* (2019) considers the importance of oral tradition in modern societies to reflect on nuclear waste deposition and how best to warn future generations of the danger beneath. Folklore and fiction mesh to become a hybrid soup of old wives' tales, natural science and Gothic phantasmagoria. Be careful when you traverse bogs and moors—many a poor soul has lost their lives to drowning and hypothermia, devilish creatures and unaccounted disappearances. Only a fool with a deathwish goes courting with a bare head on the Yorkshire uplands as detailed in 'On Ilkla Moor Baht 'at'; the will-o'-the-wisp will tempt an unwary traveller into the pitiless jaws of the mire; boggarts, selkies, imps, sprites, the king under the sea, lobs and shucks all coupie down in wet woods, ebbing tides and swampy ground, waiting for an unwary tramper to hitch a lift with, take to the underworld or, at the very least, to give them a jolly good fright.

As we explore wetlands as literary spaces, we must also consider the ways in which literature which is linked to landscape has been used to underpin certain nationalistic ideals. There is not the space within this short book to explore this in depth. Rather, we wish to call to the reader's attention that notions of 'Albion' and pastoral visions of pre-modern, untouched landscape can be accused of drawing upon jingoistic, 'völkisch' affirmations of national identity linked with purity and pedigree. Wetlands are often the spaces of aliens, foreigners, outsiders. As we consider the way in which folk tales, modern literature and other artistic forms create atmospheres around landscape, we must also use our critical faculties to interrogate the interests that may sit behind these otherwise innocuous presentations of space.

Wetlands as spaces of folk tales also accomodate changes in literary appetites. Moving from oral traditions to text based diatribes and drama, these waterscapes continue to act as settings for abject warnings over personal morality. As the modern novel holds sway as the literary form *du jour* of expounding upon and solidifying dominant social norms, customs and acceptable behaviours wetlands find their new place in contemporary culture. Jane Austen's impetuous Marianne Dashwood slips and sprains her ankle in the rain and catches a near fatal cold, and attraction, when recklessly tramping the Wiltshire heath without a male companion in *Sense and Sensibility* (2004); the barrows of Somerset's Exmoor are the backdrops to bloodthirsty feudal revenge in *Lorna Doone* (1890); the Famous

Five find themselves in a pre-adolescent Devonshire pickle with itinerants and mysterious foreigners on *Five Go to Mystery Moor* (1954).

As we forward towards more contemporary literature, fens, carrs and bogs still hold a dark and strange allure. Here the Gothic horror of spaces outside of urbanity enables all forms of dark desire to emerge. Sirens lure young men to early deaths in Daisy Johnson's short story 'Blood Rites' (2016) based in the East Anglian Fens; strange and occult goings-on occur in the inter-tidal wetlands of the Cumbrian coastline in *The Loney* (2016); a winged leviathan lolls in the fecund Blackwater estuary in the Victorian era fable *The Essex Serpent* (2017).

English literature often depicts wetlands as places of othering. These representations cast rivers, streams, ponds, bogs, moors, mires, lakes and fens as places in which the unpredictable can occur. They are liminal spaces—not quite land, or water, or air. A kind of earth magic inhabits these spaces of the imagination. Wetlands' physical shape shifting is matched by a flickering between the real and the fictional. These tropes seem to bring together the physicality of wetlands in an ecological sense, the ways in which the ecotones of wetlands create no sharp boundary point between land and water, with the atmospheres found in these spaces. Shifting light, scattered cloud caught in the peripheral gaze, wetlands are blurred, porous spaces in which each assemblage of ever-changing element, water, light, chemical process, biotic and abiotic co-production, is in motion. Attention must be paid to each wetland system and each wetland form.

What does it mean to *know* a wetland? Consideration must be given to the ways in which we have enculturated expectations and presumptions of our English wetlands. Literature has long colluded with wetlands to present them as othering spaces—places of mystery, the occult, huissiers to the underworld. By casting wetlands as delinquent spaces, those who traverse these spaces become vulnerable to temptations and to harm. Wetlands are the folkloric and fictional spaces in which ne'er-do-wells and lovers cross paths, often with negative outcomes. Pirates, buccaneers and smugglers inhabit the creeks and inlets of coastal wetlands; illicit distillers of gin, brandy, beer and perry hide stills in dilapidated bothies awaiting them; escaped convicts, turncoats, religious and political dissidents plot subversion; poachers and appropriators of others' goods hide their bounty, along with those hiding from the law, all wallowing in this submerged landscape; star-crossed lovers and careful adulterers find these as spaces of freedom from the shame of social opprobrium. Newspaper reports and historical documents attest to foul deeds committed, old and new. Folk songs, mummers plays, country lore and fairy stories all place wetlands as places that gentle folk would not tread without good reason. Yet thinking about the activities in these spaces, we can consider how much these collective social representations were less about keeping communities safe and more about keeping these areas free as non-conformist spaces. Heterotopias abound in these spaces; a multitude of possibilities can occur. As outlined in Chap. 2, on these sites people could actively continue traditional livelihoods outside of sweeping reformist moves to develop a Baconian dominancy over nature through drainage, hedging, road building, land management—all forms of enclosure successfully implemented in other landscapes, but which has always proved harder to penetrate within wetland systems.

Classical literature utilising wetlands could be then viewed as a means to continue this 'othering' of the space—to keep it away from the slow death creep of modernisation. By reflecting back to the civility and order found in mainstream society, and its ordered and highly managed landforms, literature either wittingly or unwittingly enabled long serving messages emphasising the immorality of wetlandscapes, with the effect of keeping them pristine for longer. As long as wetlands remained off kilter from mainstream values and retained their sense of otherness, and in many ways were viewed as valueless, they were left alone. In places where folk lore and literature connived to enhance the periphery nature of these spaces, such as the Fens, the moors, the Levels, wetlands were not drained to provide agricultural land to ultimately raise taxes in support of colonially driven oppressive wars; local ways of life were kept; vernacular ways of living with the ebb and flow of wetland life remained alive; pagan spirituality lasted longer in wetland English environments than anywhere else, with moon cults thriving into the eighteenth century (Hutton 2019). We can argue that wetlands are heterotopic spaces, where resistance to dominant power relations are enacted. For Foucault, these spaces may look like others; a wetland may look like a stream riven wood or salt marshes or a lake's foreshore—but they are interpreted or resonate differently (1984). It is culture that makes this difference—our interpretations and cultural representations of these different geographies that warp the space. As Johnson suggests (2006: 84):

Heterotopias draw us out of ourselves in peculiar ways; they display and inaugurate a difference and challenge the space in which we may feel at home. These emplacements exist out of step and meddle with our sense of interiority.

This reflection of wetlands as heterotopias—spaces in which to enable reactionary or revolutionary thoughts and deeds—has been taken up by literary critics. Borlik (2013) argues that many Elizabethan writers baulked at the creeping extension of monarchical reformations and land enclosure through their work. Borlik suggests that Shakespeare's Caliban is less a representative of colonial slavery and more a direct comment on the catastrophic draining and carving up of fenlands, particularly in East Anglia where this large-scale land management upheaval was led by Dutch engineers. Given Caliban's wilderness dwelling in bogs and marshes, it is suggested that it is the folklore of fen spirits which inflect the character of Caliban. Caliban is a child of nature, not a savage. His affinity with his natural surroundings is to be celebrated, rather than viewed as woe begotten. The belief in fen magic, of fen sprites and of living with the fen's flood pulses, not trying to sculpt it with science, is epitomised by the character of the Tiddy Mun (see Image 4.2).

Marie Clothilde Balfour's folklore ethnography in Lincolnshire of the Tiddy Mun (1891) is an enigmatic piece of writing, capturing communities on the cusp of transition into the modern. The Tiddy Mun was a fenland spirit, who would wreak a terrible revenge on you if you did not pay him his dues and respect the wetlands. As a means to enacting social compliance, and enabling an anti-drainage vanguard, Tiddy Mun was your go-to watery demon, a not always benign spirit of place. Drowned cattle, lost children and flooded homes would all be enchantments for a lack of accommodation with the Tiddy Mun. Borlik (2013) suggests that a grotesque genius loci can sometimes act as a kind of Bakhtinian mascot of the populace, a figure of resistance. In Lincolnshire folklore, the fen spirits would not take kindly to efforts to evict them through drainage (Borlik 2013: 24). To bolster the tale's authenticity, Balfour transcribes it in the first person as the old woman told it, complete with a phonetic approximation of her Lincolnshire dialect:

> For thee know'st, Tiddy Mun dwelt in tha watter-holes doun deep i'tha green still watter, an' a comed out nobbut of evens, whan tha mists rose. Than comed crappelin out itha darklins, limpelty lobelty, like a dearie wee au'd gran'ther. Avi'lang white hair, an' a lang white beardie, all cotted an' tangled together; limpelty-lobelty, an' a gowned i'gray, while tha could scarce see un thruff tha mist, an 'a come wi'a sound of rinnin' watter, an 'a sough o' wind, an' laughin' like tha pyewipe screech. (1891: 150–51)

4 WETLANDS AS LITERARY SPACES: OFF KILTER, OFF GRID, OFF THE WALL   101

Image 4.2 The fearful Tiddy Mun. (Credit: Benjamin Gearey)

Only by calling out to the Tiddy Mun a grovelling apology until you heard the call of a lapwing could you be assured of a safe passage through the night against rising waters. Linking back to European folklore and fairy stories, Tiddy Mun is almost an amalgam of *Rumpelstiltskin* and *Whuppity Stoorie*, unmarried males living in the wilderness that need to be appeased by various means. Explored later in the chapter, this crossover between classical literature, fairy tale and folklore forms part of another strand of literature: the English Eerie. It is enough to say here that to live a wetland life was to inhabit a very particular place in society, one that marked out a resident, and their kin, as otherworldly. Unlike those who lived in cosy villages in the hills, or in coastal cottages facing the sea, wetland peoples were depicted in folk tales as insular, suspicious, uneducated.

Those who lived in the wetlands were known as 'wetlanders, as the residents of the Fens came to be known in the medieval and early modern periods' (Noetzel 2014: 4). They were therefore distinct from uplanders or foresters:

> One of the most foundational elements of this narrative is the pervasively uncanny nature of wetland landscapes and their definition as the Other to the centres of English culture that have been present since the medieval period. This monstrous otherness manifests itself throughout the history of the Fens, beginning with the Anglo-Saxon understanding of this landscape as the realm of oath-breaking and outlawed humans, monsters, and demons. (Noetzel 2014: 9)

For now we can say that the wetlands of our classical cultural imagination generate a cultural frisson. They are push-pull places, where social transgressions can occur. Very much like the benighted villagers who live outside of the castle of Mervyn Peake's *Gormenghast* trilogy (1999), their surroundings have addled their minds. These wetlands are crepuscular spaces of danger and foreboding, where the flatness and the vastness of the moors, bogs, mires and heath can disorientate with no easy landmarkings. Woods and holloways cast shifting shadows that play tricks on the eye. Roaring coasts offer treasures from the deep and off-comers riding the tides. Wetlands then have a dark romance, a gravitational pull which is reflected in literature from other cultures too. Bram Stoker's *Dracula* arrived at Whitby by boat (1965); H.P. Lovecraft's *Cthulhu* (2014), a terror from the deep, uses subterranean networks to travel unseen; the Slavic water nymphs, the Rusalkas, emerge from rivers in spring under the full moon to lure young men to join them in their crystalline aqueous palaces. As we see in Image 4.3, the Rusalkas were bewitching entities, slim and naked, their modesty covered by long white tresses that glowed in the moonlight. As sites for an illicit rendezvous, the forested lakes offer space for transgression: whether for poaching, wildfowling, plotting or meeting unsuitable lovers. As the depiction indicates, the Rusalkas offer unnamed temptations. Yet, look past their alluring smile and you would see dead eyes and long, strong fingers ready to grasp you and pull you under the cold dark waters within the forest.

Water then provides the perfect medium for an altered state of reality. Moving between liquid, visible gas and solid, the water on the ground of the wetlands mingles with the sky and the land to enable transformative states of sensorial experience. Many writers argue this is an essential

Image 4.3 Rusalkas. (Credit: theweirdandtheodd.com)

element to what makes us human—our deep connectivity with water in all its forms, for its haptic and somatic sensory pleasure giving. Water environments are also cognitively formational to our sense of place—once seen, felt or touched, we inhabit these landscapes again in our mind's eye. This is explored in a multimedia piece of literature: the poet Simon Armitage's petroglyphs on six standing stones, spiked brontosaurus like on the tops of the Pennine's watershed. Here, along 47 miles of wetland from Marsden to Ilkley Moor, the Stanza Stones are etched by stonemasons Pip Hall and Wayne Hart with water poems (2013). Each poem is orientated around the six forms of water found high on these moors: Rain, Mist, Snow, Dew, Beck and Puddle. To visit these poems, the reader must move between these elemental states, becoming suffused with the water, to then find a final encounter in visual text. Armitage and the stonemasons ask the walker-reader to really connect with the wetlands they are

Image 4.4 The Beck Stone; part of The Stanza Stones Walk, West Yorkshire and Greater Manchester, UK. (Credit: Mick Melvin)

traversing—the poet affirming the wetlands as both a physical and literary space. Recalling Bachelard's work *The Poetics of Space* (1958) and its call to provide more consideration towards the phenomenological experience of built environments and nature, we can see these Stanza Stones monoliths are Armitage inviting the walker-reader to dissolve into the wetlands. As Bachelard puts it: 'the poet speaks on the threshold of being' (Bachelard 2014: xvi). This close connectivity with our water landscapes, the altered reality they provide, is a dominant theme within a different branch of literature which informs our understanding and appreciation of wetlands: nature writing (Image 4.4).

Nature Writing: The 'Lone Enraptured Male' and forays into Ecocritical Theory

Kathleen Jamie's three-word summary of contemporary nature writing, the work of the 'lone enraptured male' (2000), has a double sting. It also describes classic or established nature writing (ENW)—the habitus of Aldo Leopold, Henry Thoreau, Gary Snyder and Patrick Leigh Fermor, amongst a canon of established names. Nothing much seems to have changed; or has it?

ENW as a genre has its roots in the age of the Renaissance, where poetry, naturalism, science and arch rationalism combine to present a different set of relationships with nature, with landscape and with ourselves. This is the era of the sublime, where landscapes once feared and loathed became both desired spectacle and revered, within an almost animist cultural milieu, celebrating the dominant power of nature over mankind. Wordsworth has been described by Tim Fulton (2011: 23) as the first 'Romantic ecologist' for his ability to connect the utter dependence of humans on nature and, in particular, on the water cycle. Wordsworth gives full consideration to hydrology and the emergence of springs and rivers from deep underground (Fulton 2011: 23):

> *PURE Element of Waters! Wheresoe'er*
> *Thou dost forsake thy subterranean haunts,*
> *Green herbs, bright flowers, and berry-bearing plants,*
> *Start into life and in thy train appear:*
> (*The Miscellaneous Poems of William Wordsworth* part 2, entitled 'Pure Elements of Waters!' p. 161)

This embodiment of nature in ENW, the close connectivity between humans and their environments, is shaped by the language used to describe landscape. Increasingly writers spoke of being awestruck, dumbfounded, rendered mute by the ingeniousness and immutability of nature. Science, particularly botany, geology and entomology, rapidly widened our understanding of the natural world in the late 19th and early 20th century. A growing appreciation of the significance of human impacts on landscape developed from what we might now consider auto-ethnographies of writers in landscapes familiar to them. Aldo Leopold wrote of a 'land ethic' in his *Sand County Almanac* in 1949; Gary Snyder's Deep Ecology perspectives were drawn from his poetry developed as a fire tower guard over long, dry summers; Roger Deakin's abundant love for his Walnut Tree Farm and surrounding waterscapes. These writers view their connectivity to their local landscapes as an ecologist views a wetland as an ecotone space—ecologists see no boundary in wetlands, only a transmutation between wet and dry ecosystems. So this concept can also be used to describe literary blurrings of imagination drawn from wetland spaces. A more recent generation of writers now champion wetland preservation.

Some are scientists, like Edward O. Wilson, who writes about the marshes and swamps of Georgia and Florida, where he learned the principles of biodiversity and biogeography. Others are reporters and essayists like Barry Lopez, Peter Matthiessen and Bill McKibben, who consistently portray American wetlands as centres of ecological harmony and rallying points for environmental activists.

A debt of gratitude is owed to scholars of post-processual archaeology (McGlade 2004; Hodder 2004). Thanks to them, and alongside human geography cohorts such as Doreen Massey and David Matless, when we consider humans and landscape across time, we reflect upon the subjectivity and contextuality of the interpretations that we make. Archaeologists enable us to read the inscriptions on landscapes left by predecessors. Land shaped by human agency over time outside of history, somewhere between antiquity and the portal towards deep time, tells us something quintessential about our changing relationships with the natural world we inhabit. Contemplating burial chambers, cleared forests, oak-tracked wetlands, diverted rivers, peat clearances and more, all serve to remind us that the way we use, access and consume natural resources is fragile, liable to change. The 'mark makings' (Ingold 1993) inscribed on our English landscapes attest to our changing relationships with our environments. These relationships are constantly changing and evolving.

We can also apply this to reflections upon the types of literature that address the different arenas of human, more-than-human and landscape interaction. Literature shapes how we see the world and is shaped itself by the culture(s) it is immersed within. A phenomenological interpretative analysis framing helps us to consider the impact of writing on how we view and value natures. Our appreciation or disdain for natures is reflected in the literature we choose to venerate. Understanding this enables us to see connectivities between different types of literature concerned with humans and landscape, and the evolution of these works. Rather than see them as literary silos, we can use interpretivist critiques to see forms or ways of writing about landscape as historically and socially constructed, and that there are transitions between these genres which are not uni-directional, but move backwards and forwards as cultural tastes change. Moreover, many pieces of literature can easily sit in multiple categories. The work of author Alan Garner can at any one time be bracketed as children's literature, as epitomising the 1970s folk horror oeuvre in writing, plays, televised dramas and radio series and as connecting with theological classical fiction such as that of C.S. Lewis and J.R.R. Tolkien. So when we think about wetlands as literary spaces, we must appreciate that like the landscapes they are drawn to, these works will sit across categories and mutually influence each other across genres and across time.

The 'new nature writing' (NNW) movement embodies this porous, reflective and forward casting set of characteristics. It's almost impossible to definitively say a time, an author or a piece of work which defines NNW, and our interests here are specifically those works that are particularly drawn to English wetland environments. It seems tautological to state it, but if there is a 'new', there must be an 'old' or at least established nature writing (ENW). Boundaries blur, but for brevity it may be better to state a claim as to what ENW represented and how NNW has built on this oeuvre, particularly with regard to wetlands and waterscapes.

ENW is not botanically inclined natural history. That is a distinct, scientifically driven form of communicative, pedagogical literature, which is orientated around certain forms of documentation and analysis for a range of readers. However, at the turn of the Twentieth century, this more scientific literature spawned offshoots which developed into more stylised forms of reflections on personhood in relation to natural surroundings. The works of Patrick Leigh Fermor and Laurie Lee, W.G. Hoskins' *The Making of the English Landscape* (1955), H.C. Darby's 1956 work *The Draining of the Fens*, Alfred Watkins' *The Old Straight Track* (2014) and Nan Shepherd's *The Living Mountain* (2008) are all descendants of sorts from Henry Thoreau, Mark Twain, Wendell Berry, John Ruskin and John Muir, amongst others who felt drawn to write about human connectivity to landscape. This humanist perspective sought in many ways to overcome the Cartesian split between science and nature, emotions and rationality. These writers, in all their different ways, perceived that humans are shaped by the landscapes they inhabit. Explorations develop either through self-analysis and self-reflection, much like Jack Kerouac and Gary Snyder's work on the loneliness and absorption of forest fireguarding in isolated look-out towers, or through the travelogues of Laurie Lee and Patrick Leigh Fermor and their interactions with locals in situ in different geographical locations and landscapes. Kerouac's *Desolation Angels* is an exemplary work of an almost meditative absorption into nature. Formative across the ENW genre is that place matters and that a sense of place is what shapes a person, and—more broadly—a community.

Throughout the 1960s and 1970s, ENW becomes saturated with works which develop an overtly political dimension. Rachel Carson's *Silent Spring* (1965) attests to this, as does Snyder's work as it sets out, along with Arne Naess, an agenda for forms of 'Deep Ecology'. There is no space in this chapter to document the wider impacts, but it is enough to say that waterscapes, and work within which ENW specifically addresses wetlands, are subsumed within the call to arms of the environmental movement. Quite rightly more pressing concerns become the focus of scholarly attention.

The renaissance of a focus on particular landscape forms is a central feature of NNW or—more precisely—British NNW, BNNW. For many Roger Deakins' body of work, with *Waterlog* (2011) and *Wildwood* (2007), as emblematic of a beginning of a new zeitgeist in nature writing. Deakins places himself as the narrator and central character in his nonfiction work, as his considerations of the world are generated through paying close attention to his immediate surroundings and his own particular interests. What is different from ENW here is that BNNW from this point on does not attempt to portray the writer as knowledgeable across disciplines, but as intense, almost obsessive in their interests. Deakins sets out in *Waterlog* to cross landscapes only through immersion in water bodies, inspired by his love of swimming in his moat at home in Norfolk, almost in a rural homage to John Cheever's *The Swimmer*. Deakins' love of cold water swimming has since inspired a raft of meditations on the human body suspended in different watery realms. Olivia Laing (2016) sets out to swim the Ouse in an ode to Virginia Woolf as documented in *To the River*; Alice Oswald's *Dart* (2011) imagines the River Dart as a mobile living organism, much like a deer or a hare, racing across the landscape with her entry into the water, a form of transmutability into a different, spirit animal self. Francis Pryor's *The Fens* (2019) is an immersive exploration into both the vast fen landscape and the life journey of the author. Jean Sprackland's *Strands* (2012) bestrides land and sea in the intertidal wetlands of her home in Lancashire and offers a year of findings and uncoverings in the fluid liminal zone between self and nature. Adam Nicolson's 1986 piece *Wetland: Life in the Somerset Levels* is instructive of this earlier movement towards forms of landscape ethnography. Julia Blackburn's *Time Song* (2019), as discussed in depth in Chap. 5, provides an imaginative immersion into shifting landscapes, asking us to consider what it means to be human. Pain memoirs also become a sub-genre of BNNW connected with water; Katherine Norbury's (2015) *Fish Ladder* explores a metaphysical swimming upcurrent across England, coping with loss and restoring a new sense of self. Psychogeography takes to the water in Rachel Lichtenstein's (2016) *Estuary*, using the Thames outflow to consider changing lives and landscapes. There are plenty of other examples of English waterscapes in BNNW; Charles Rangeley-Wilson's *Silt Road* (2013), Patrick Barkham's (2015) *Coastlines*, and the blog site *Caught By the River's* collected works edited by Jeff Barnett are a submergent foray into the special place that rivers have within our human hearts.

No excursion into NNW can take place without recognition being given to the colussi of the genre, those whose work moves outside of the close attention to personhood that marks many of the works. These writers' ambits span multiple published works: Robert Macfarlane, Richard Mabey, Mark Cocker, Kathleen Jamie, Jim Crace, Simon Armitage, Ted Hughes, Julia Blackburn, Simon Schama and those who have or are working outside the UK; Tim Robinson, Barry Lopez, Seamus Heaney, Amitav Ghosh, Rebecca Solnit, Carolyn Finney, Bill McKibben, Annie Dillard, Major Jackson, Janice Harrington and Mahesh Rangarajan, amongst others. These writers, working across disciplines, attracting readers through published words, public events and open readings, podcasts, Twitter feeds and lectures, all attest to the vibrancy of thinking through landscape.

Folklore, folk tales, traditional songs, ghost stories and tales of the fantastical are all well-examined aspects of English culture. Rogues, luckless lads, summertime courting, heartbreak and derring-do are prominent narratives within these stories. Landscapes play an intimate role in shaping their story arc, with sea shanties to advise against setting sail in bad weather, and wooded vales the sites for clandestine meetings and ill-advised romances. Wetlands are one such landscape—misty moors for getting lost and found, rivers for drowning and for dreaming, coasts for smugglers and pirates, lakes for losing and retrieving artefacts of antiquity, as well as losing loved ones.

The English Eerie is predicated on the frisson of the unknown in landscapes close to home, on the borders or the boundaries of the domestic. Tolkien's Middle Earth has this realm of the unknown only a day's ride away—tantalisingly close, particularly for those closely bounded to the earth. Farmers, plough hands, milkmaids and wood gatherers inhabit these spaces in the folkish imagination. Landscape is a place of refuge and sustenance as well as holding the potential to deliver sinister experiences with the rise of the mist or the fall of the sun. Freud's examination of the *unheimlich*, which Mark Fisher (2016) argues is not so much concerned with the 'uncanny' as with the 'unhomely', conveys a growing sense of disease which we can experience in certain landscapes. Wetlands become representational of these spaces: barren moors with no clear features to

Image 4.5 Bungay's infamous Black Shuck town spire, Bungay, Suffolk, UK. (Credit: The Suffolk Coast DMO)

quickly disorient the walker; wet woodlands hung with sopping moss and infernal tangles of trees and ankle-breaking roots; mosquito-infested mires, with bogs like quicksand ready to swallow an unsuspecting interloper.

English folk stories are riven with the weird and eerie landscape of the wetland. Fairy stories are the most obvious entry route as they enmesh with folk stories: the selkie stories of mermaids and mermen from all around coastal England; Arthur and the sword Excalibur retrieved from the Lady of the Lake; the tale of the West Country Moonrakers, smugglers pretending to confuse the moon with a wheel of cheese to mask their submerged contraband from the excise men, the black shuck which prowls the moors of Norfolk, Suffolk, Yorkshire (Image 4.5).

Libidinous Zombies, Swamp Monsters, Urban Fantasy and Cli-Fi

As much as wetlands are erroneous spaces—places in which to both physically stray and mentally wander across complex terrains—they are erroneous spaces in literature too. As we have seen, they provide the imaginary architecture for all manner of storytelling to take place. In the peripheral vision of the English Eerie sits other sub-genres, hidden in plain sight. These other wetland literatures reflect back to us both changes in modern publishing and improved accessibility of writing for new authors. They also remind us of the crossover of modern science fiction writing with a form of future politics that has always been an elemental part of this particular genre.

Electronic publishing (e-publishing) has enabled a new generation of writers and readers to flourish. Although physical book sales continue to decline (*The Guardian*, 25 June 2019), the use of electronic reading devices and other virtual ways of assimilating non-visual stories—Kindles, e-readers, phone apps, tablets, podcasts and audio books—continues to rise. Virtual reality headsets to meld with online gaming are technologies of the new future.

One aspect of this new oeuvre is its accessibility across all reading sectors and particularly an appropriation of the punk ethos of 1970s fanzines to e-publishing. Increasingly fan-fictions, in other words, members of subcultural groups such as Straight Edge (Haenfler 2004), Otaku (Okamoto 2015), Seapunk (Longhurst et al. 2017), and post-human rap (Adams Burton 2017), amongst others, produce their own writing for consumption amongst discrete groups. The speed, ease and affordability of e-publishing have enabled everyone to become an author. As a result there is a boom in e-publishing sub-genres, including bizarro fiction, domestic noir and climate change science fiction or 'cli-fi', amongst many others (Murphy 2017), together with elements which straddle traditional literary markets such as the new weird (Noys and Murphy 2016), slipstream (Broderick 2000) and ero-guru (Paasonen 2010). Image 4.6 is representative of this new e-publishing bizarro fiction. As we see from the cover, it utilises steampunk/*Mad Max* dystopian aesthetics alongside a firmly tongue-in-cheek nod to 1950s Hollywood B-movies. The femme fatale is not a bewitching alien, as mooted by Dr John Cooper Clarke's poem *(I married a) monster from outer space*, but a semi-clad she-wolverine with a lascivious pout and a low slung belt of ammunition for her pump action shotgun. *Warrior Wolf Women of the Wasteland* is an about-turn to Louisa M. Alcott's vision of female capability. Is the wasteland their mutant inheritance, or have they created it? An allegory for our times we fear.

Wetlands as 'othering' spaces move in and out of focus amongst these alternative, and much loved, works and authors. China Miéville's work encompasses new weird, cli-fi, slipstream, postmodern philosophy and (cyber)punk ecology, doffing a self-knowing trilby to H.P. Lovecraft, Le Guin and M.R. James, with emblematic works with wetlands as instrumental landscapes in books such as *The Tain* (2002) in which the River Thames and its marshy surrounds play as both refuge and hunting ground, *Kraken* (2010) in which London's flooded subterranea is the site of a giant octopus death cult and *The Scar* (2002) in which steampunk pirates infiltrate creeks and rivers to loot and power-play. Across many of these

Image 4.6 Warrior Wolf Women of the Wasteland. (Credit: Carlton Mellic III)

works, the unpleasant consequences of changing climates lurk as reminders of how our own lives are subject to climate change-linked anxieties. The popularity of this form of imaginative or speculative fiction seems to have emboldened a new generation of writers. The writing of Mark Fisher utilises the '*weird and the eerie*' (2016) English landscapes of the moors, the heath and the dank woods to capture the disorientating feeling of unloved wetland spaces. Fisher connects this dis-ease with the disconnection from landscape wrought by late capitalism, as does the Derridean 'hauntology' of British pastoral noir as described by Adam Scovell (2017), and epitomised in the cult horror screenplays which utilise English wetlands within the imagined worlds of Ben Wheatley (*A Field in England*), Nigel Kneale (*The Stone Tape*), Anthony Shaffer (*The Wicker Man*—although Summer Isle is not nationally defined) and Alan Clarke (*Penda's Fen*).

Of all the fiction writing, old and new, which uses English wetlands as a backdrop for othering, whether these are blood sucking, body deposition or doomed romances, climate change science fiction, or cli-fi, is the area that has a growing readership across genres, moving firmly into the mainstream. These books reimagine landscapes as living beings, reacting to the pestilence and degradation they are subjected to by humans within the Anthropocene with grotesque and sinister responses. J.G. Ballard's *The Drowned World* (2012) is viewed as the benchmark for this oeuvre; David Mitchell's (2014) *The Bone Clocks*, Benjamin Warner's (2016) *Thirst* and Christopher Lambert's *Tales from the Black Meadow* reimagine beings from English waterscapes seeking revenge on complicit humans. Even the psychogeographers can be deemed to dabble with cli-fi such as Nick Papadimitriou's *Scarp* (2012) and *Marshland: Dreams and Nightmares on the Edge of London* by Gareth E. Rees (2013).

These works call our attention to the diversity of academic texts that also draw upon these different presentations of wetlands, to ask us to pay more attention to their role in creating modern societies. Amongst these writers, the work of Rod Giblett has been most influential. Expanding outwards from his ethnographical experiences in Australia with his local swamp, covered in *Black Swan* (2013), Giblett has written extensively on the forming of modern cities, such as London, New York and Chicago, from swampland. A hugely influential work, *Postmodern Wetlands* (1996), attests to the 'quaking zone' of unstable land we have only temporarily reclaimed from nature without realising the long-term impacts of our desire to have mastery over nature. Giblett's work also makes reference to other academic writing which draws on the natural sciences but which inevitably leads to more philosophical reflections on our relationship with nature. From an English point of view, we can also draw upon W.G. Hoskins' and Watkins' work and the poetry of John Clare who Rod Giblett has anointed as the 'patron saint of the Fens' (Giblett 2019). Carol Donaldson's *On the Marshes* (2017) embeds us in the Kent edgelands. William Atkins' *The Moor* (2014) perambulates the reader through mire and grassy tussocks. To conclude, we see that whilst there is a diversity of literature which draws upon wetland landscapes, what connects them is their treatment of these environments as quintessentially 'othering' spaces. Not back gardens, public parks, beaches or downlands. Wetlands hold their own peculiar and magical charms, but can also be perceived as places of anxiety and threat. As we move into the next chapter, we see the ways in which wetlands have long been held in regard by humans as othering spaces—places in which to form, retrieve and enact memory making in a variety of guises.

References

Adams Burton, J. (2017). *Posthuman Rap*. New York: Oxford University Press.
Armitage, S., Hall, P., & Lonsdale, T. (2013). *Stanza Stones*. London: Enitharmon Press.
Atkins, W. (2014). *The Moor: Lives Landscape Literature*. London: Faber & Faber.
Austen, J. (2004). *Sense and sensibility*. Oxford: Oxford University Press.
Bachelard, G. (2014). *The poetics of space*. London: Penguin Classics.
Balfour, M. C. (1891). Legends of the Cars. *Folklore, 2*(2), 145–170.
Ballard, J. G. (2012). *The Drowned World: A Novel*. London: WW Norton & Company.
Barkham, P. (2015). *Coastlines: The Story of Our Shore*. London: Granta Books.
Blake, W. (1994). *Songs of Innocence and of Experience* (Vol. 5). Princeton: Princeton University Press.
Blackmore, R. D. (1890) Lorna Doone: a romance of Exmoor (No. 666). London: Harper & brothers.
Blackburn, J. (2019). *Time song: searching for Doggerland*. London: Jonathan Cape.
Blyton, E. (1954). *Five Go to Mystery Moor*. London: Hodder and Stoughton.
Borlik, T. A. (2013). Caliban and the fen demons of Lincolnshire: the Englishness of Shakespeare's Tempest. *Shakespeare, 9*(1), 21–51.
Broderick, D. (2000). *Transrealist Fiction: Writing in the Slipstream of Science* (Vol. 90). London: Greenwood Publishing Group.
Brontë, E. (2009). *Wuthering heights*. Oxford: Oxford University Press.
Brown, T. (1958). The black dog. *Folklore, 69*(3), 175–192.
Bunyan, J. (1967). *The pilgrim's progress*. Grand Rapids: Zondervan Publishing House.
Carson, R. (2002). *Silent spring*. New York, NY: Houghton Mifflin Harcourt.
Cheever, J. (1978). *The Stories of John Cheever*. New York: Knopf.
Colebrook, M. (2013). Comrades in Tentacles: HP Lovecraft and China Miéville. In D. Simmons (Ed.), *New Critical Essays on HP Lovecraft* (pp. 209–226). New York: Palgrave Macmillan.

Darby, H. C. (2011). *The draining of the Fens*. Cambridge: Cambridge University Press.
Deakin, R. (2007). *Wildwood*. London: Hamish Hamilton.
Deakin, R. (2011). *Waterlog*. London: Random House.
Du Maurier, D. (1936). *Jamaica Inn*. London: Little Brown.
Dickens, C. (1993). *Great expectations* (Vol. 8). Oxford: Oxford University Press on Demand.
Donaldson, C. (2017). *On the marshes: A journey into England's waterlands*. Dorchester: Little Toller.
Fisher, M. (2016). The weird and the eerie. *London: repeater books*.
Fort, T. (2008). *Downstream*. London: Random House.
Foucault, M. (1984). *The Foucault reader*. London: Pantheon.
Giblett, R. J. (1996). *Postmodern wetlands: culture, history, ecology*. Edinburgh: Edinburgh University Press.
Giblett, R. J. (2013). *Black Swan Lake: life of a wetland*. Bristol, UK: Intellect Books.
Giblett (2019) Personal communication, 25th September 2019.
Gill, C. (2013) Blackstone vampire series, on Kindle.
Gurr, J. M. (2010). Emplotting an Ecosystem: Amitav Ghosh's The Hungry Tide and the Question of Form in Ecocriticism. *Cross/Cultures, 121*, 69.
Haenfler, R. (2004). Rethinking subcultural resistance: Core values of the straight edge movement. *Journal of contemporary ethnography, 33*(4), 406–436.
Heaney, S. (2009). *Beowulf*. London: Faber & Faber.
Hodder, I. (2004). *Theory and practice in archaeology*. London: Routledge.
Hoskins, W. G. (1955). *The Making of the English Landscape*. London: Hodder and Stoughton.
Hurley, A. M. (2016). *The Loney*. London: John Murray Press.
Hutton, R. (2019). *The triumph of the moon: A history of modern pagan witchcraft*. New York: Oxford University Press, USA.
Ingold, T. (2003). *Key debates in anthropology*. London: Routledge.
Jamie, K. (2008). A lone enraptured male. *London Review of Books, 30*(5), 25–27.
Johnson, D. (2016). *Fen*. London: Random House.
Johnson, P. (2006). Unravelling Foucault's 'different spaces'. *History of the human sciences, 19*(4), 75–90.
Jonson, B. (1994). The devil is an ass. In P. Happé (Ed.), *The Revels Plays*. Manchester: Manchester University Press.
Laing, O. (2011). *To the River: A Journey Beneath the Surface*. London: Canongate Books.
Lambert, C. (2013). *Tales from the black meadow*. London: Exiled.
Leopold, A. (1949). *A Sand County almanac*. Oxford and New York: Oxford University Press, USA.
Lichtenstein, R. (2016). *Estuary: Out from London to the Sea*. London: Penguin UK.
Longhurst, B., et al. (2017). *Introducing cultural studies*. London: Routledge.

Lovecraft, H. P. (2014). *The call of Cthulhu*. London: Simon and Schuster.
Macfarlane, R. (2012). *The old ways: A journey on foot*. London: Hamish Hamilton.
Macfarlane, R. (2019) *Underland: A Deep Time Journey*. Penguin UK.
McGlade, J. (2004). The times of history: archaeology, narrative and non-linear causality. In T. Murray (Ed.), *Time and archaeology* (pp. 15–34). New York: Routledge.
Miéville, C. (2002a). *The scar*. New York: Del Rey.
Miéville, C. (2002b). *The tain*. London: PS Publishing.
Miéville, C. (2010). *Kraken*. Basingstoke: Pan Macmillan.
Mitchell, D. (2014). *The bone clocks*. New York: Random House.
Muir, LL (2015) The ghosts of Culloden Moor series (kindle series). https://www.amazon.com/gp/product/B01049Y714?notRedirectToSDP=1&ref_=dbs_mng_calw_0&storeType=ebooks
Murphy, B. M. (2017). *Key concepts in contemporary popular fiction*. Edinburgh: Edinburgh University Press.
Noetzel, J. T. (2014). *Marsh men and trackless bogs: A cultural history of the English fens*. Saint Louis: Saint Louis University.
Norbury, K. (2015). *The Fish Ladder: a journey upstream*. London: Bloomsbury Publishing.
Noys, B., & Murphy, T. S. (2016). Introduction: Old and new weird. *Genre*, *49*(2), 117–134.
Okamoto, T. (2015). Otaku tourism and anime pilgrimage phenomenon in Japan. *Japan Forum*, *27*, 12–36.
Papadimitriou, N. (2012). *Scarp*. London: Hodder & Stoughton.
Paasonen, S. (2010) Repetition and hyperbole: The gendered choreographies of heteroporn. In K. Boyle (ed.) Everyday pornography. London and New York: Routledge, pp.63–76.
Peake, M. (1999). *The Gormenghast Trilogy*. London: Random House.
Perry, S. (2017). *The Essex Serpent*. New York: Custom House.
Poe, E. A. (1914). *Edgar Allan Poe's tales of mystery and imagination* (Vol. 336). London: JM Dent & Sons, Limited.
Pryor, F. (2019). *The Fens: Discovering England's ancient depths*. London: Head of Zeus.
Rangeley-Wilson, C. (2013). *Silt road: The story of a lost river*. London: Random House.
Ransome, A. (2001). *Swallows and amazons* (Vol. 1). London: Random House.
Rees, G. E. (2013). *Marshland: dreams and nightmares on the edge of London*. London: Influx Press.
Scovell, A. (2017). *Folk Horror: Hours Dreadful and Things Strange*. Leighton Buzzard, UK: Auteur Publishing.

Shakespeare, W. (1964) *The Tempest*. Edited by Robert Langbaum. New York: New American Library.
Shakespeare, W. (2005). *The Tragedy of King Lear*. Cambridge: Cambridge University Press.
Shepherd, N. (2008). *The living mountain: a celebration of the Cairngorm mountains of Scotland*. London: Canongate Books.
Sprackland, J. (2012). *Strands: A year of discoveries on the beach*. London: Random House.
Stoker, B. (1965). *Dracula*. New York: Signet.
Sutherland, P., & Nicolson, A. (1987). *Wetland: Life in the Somerset Levels Michael Joseph publishing*. London.
Tolkien, J. R. R. (2012). *The Lord of the Rings: One Volume*. London. Houghton Mifflin: Harcourt.
Warner, B. (2016). *Thirst*. London: Bloomsbury Publishing.
Watkins, A. (2014). *The old straight track*. London: Head of Zeus.
Woolley, J. (2018) Hounded out of time: Black Shuck's Lesson in the Anthropocene. Environmental Humanities, 1 0(1): 295–309.

Websites

https://theconversation.com/how-19th-century-fairy-tales-expressed-anxieties-about-ecological-devastation-73137
https://www.terriwindling.com/blog/fairy-tales/
https://www.theguardian.com/books/booksblog/2018/mar/27/ursula-k-le-guins-electronica-album-music-and-poetry-of-the-kesh
https://www.theguardian.com/books/2015/apr/10/eeriness-english-countryside-robert-macfarlane
https://www.historic-uk.com/CultureUK/The-Moonrakers/ accessed on 10th June 2019
https://www.independent.co.uk/news/long_reads/folk-realism-english-literature-countryside-legends-landscape-nature-gothic-writers-fantasy-a8234691.html
http://www.literarynorfolk.co.uk/fens.htm
https://www.independent.co.uk/news/long_reads/folk-realism-english-literature-countryside-legends-landscape-nature-gothic-writers-fantasy-a8234691.html
https://en.wikipedia.org/wiki/Slavic_water_spirits
https://strangewetlands.wordpress.com/2012/10/22/swamp-music-revisited-a-new-take-on-william-blake/

https://www.erudit.org/en/journals/ravon/2010-n57-58-ravon1824552/1006512ar/
https://books.google.co.uk/books?id=O90kAAAAMAAJ&pg=PA166&lpg=PA166&dq=PURE+Element+of+Waters!+Wheresoever+Thou+dost+forsake+thy+subterranean+haunts&source=bl&ots=V6Vfl6I00W&sig=ACfU3U3-G4NRyj-XMIBvYP9kzocjEQ_ryA&hl=en&sa=X&ved=2ahUKEwjAkI3W8_XiAhVRQkEAHSUlB10Q6AEwAHoECAgQAQ#v=onepage&q=PURE%20Element%20of%20Waters!%20Wheresoever%20Thou%20dost%20forsake%20thy%20subterranean%20haunts&f=false

CHAPTER 5

Wetlands as Remembrance Spaces: Contemplation, Ceremony and Commemoration

Abstract In this chapter we explore the importance of English wetlands to humans as places for contemplation, ceremony and commemoration. In the opening section, we provide an overview of the types of remembrance practices associated with wetlands, including tree planting events, memorial benches, bird hide plaques, as well as intimate personal performances such as poetry, artwork and scattering ashes. We explore how pre-modern uses of wetlands, as liminal threshold spaces where life and death converge, sit alongside these modern practices of remembrance. Looking at our three case study sites in detail, we explore the differing ways these spaces have been marked by human memorial practice over time. Labyrinths, turf mazes, raised wooden trackways, earthworks and henges from antiquity are considered alongside the recent past, represented by pill boxes, peat workings and military-industrial artefacts. These all lead us to examine aspects of heritage and legacy within wetland spaces and how they are shaped by the repurposing of space by neoliberal capitalism but also offer opportunities for a more civic commons.

Keywords Contemplation • Ceremony • Commemoration • Remembrance • Pre-modern • Liminal threshold • Labyrinths • Turf mazes • Heritage • Legacy

Introduction

Wetlands have long been regarded as liminal spaces, where earth, sky and water meet in fluid states of materiality (Van de Noort and O'Sullivan 2006; Tilley 2010; Fredengren 2016). As we've seen in Chap. 4, wetlands are often depicted historically in multiple cultural representations as miasmic, forbidding and uncanny. Wetlands physically sit apart from society—and those humans who have made them spaces for sustenance, recreation, ceremony or occlusion are likewise presented as existing on the margins (Giblett 1996; Coles and Coles 1986). Wetlands can be thought of as multi-use spaces whose status and value reflect each historical epoch's regard for these complex terrains. In this chapter, we take time to reflect on the importance of wetlands for humans as remembrance spaces over time. We draw upon the works of philosopher Maurice Halbwachs, geographer Yi-Fu Tuan and 'Deep Ecology' writer Arne Naess, amongst others, to explore how identity and memory are innate attributes of what makes a person connected to both their natural surroundings and to other beings, both human and the more-than-human. The importance of wetlands here are their unique attributes. A connected collective of land, water and sky, wetlands have drawn humans into very particular site-related practices over vast swathes of time. The chapter thus commences with a consideration of the perceived 'wildness' of these landscapes, as places where contemporary humans can consider deep time.

Evidence suggests that English wetlands have long been important to humans as places for contemplation, ceremony and commemoration. In the opening section, we provide an overview of the types of commemorative practices that are associated with English wetlands, to move towards considerations drawn from our research data around contemporary performances of memory and ceremony on our case study sites.

Pre-modern uses of wetlands, as liminal spaces where life and death converge, sit alongside modern practices of remembrance. Looking at our three case study sites in detail, we explore the differing ways these spaces have been marked by human memorial practice over time. In the Somerset Levels, under the watchful eye of Glastonbury Tor, Neolithic trackways abut benches with memorial plaques. In Bedfordshire the commemorative planting of arboreal copses stand above streams where twinkling diyas float downstream on holy days. In Alkborough Flats turf labyrinths from antiquity share eyelines with bird hides dedicated to past ornithologists. Making use of the different remembrance narratives of the research participants, we explore the ways in which these saturated spaces generate embodied responses of inclusion, through which the respondents detail their immersion into the landscape, becoming intimately connected to their surroundings. These memorial practices can be highly mobile physical engagements, which involve digging, painting, walking, photographing, crafting, and also more contemplative, such as sitting, reflecting, encountering, accepting. Connecting humans across time and space, wetlands can be repositioned within our cultural imaginings as important spaces of otherness, where marginality is to be championed rather than decried.

Building on this idea of engagement, the chapter considers the ways in which modern political economy formations and, in the English example, the increased marketisation and management of green-blue environments, shape how we perceive these spaces as places for memory making and as alternative forms of peripheral commemorative practices that take place on these sites. Through this we explore current ruminations of heritage and the monetisation of leisure pursuits gravitated around both reconstructive archaeology and the recent past, exemplified by contemporary theorists such as Caitlin DeSilvey and Mark Fisher. The chapter uses these insights to ask the reader to reflect upon the balance that needs to be struck between celebrating the wellbeing benefits arising from these intimate human-wetland relationships and the reductive, prescriptive manipulations through which these spaces are seen as a cheap remedy for societal ills generated by capitalism.

Wetlands Over Time

Wetlands, and in particular peat bogs, are a treasure trove for enhancing our understandings of the lives and cultures of our human forebears. The propensity towards low oxygen levels in some bogs and marshes, if teamed with acidic conditions, can make these wetlands ideal for the preservation of placed items and wooden monuments (Mitsch et al. 2009; Bond 2009). These items include raised trackways and platforms (https://ioahc.net/open-day-neolithic-trackway-platform-reconstruction/), marsh forts (Norton 2019) and crannogs (Fredengren 2016), vessels such as boats and jars (https://thepallasboyvessel.wordpress.com), jewellery and other high status items such as totems and weaponry (see the reconstructed Ballachulish goddess in Image 5.1: https://thepallasboyvessel.wordpress.com; Cowie 2013) and, most notably, bodies—of humans and their

Image 5.1 Reconstructed Ballachulish figure from the Pallasboy project, carved by Mark Griffiths. (Credit: Benjamin Gearey)

companion animals (Turner and Scaife 1995; Fredengren 2016). Some are the misplaced items of everyday wetland life from antiquity—fish traps, kegs of butter, turf cutting tools. Given the lack of material evidence available concerning cultures from 'prehistory', when organic matter has decomposed and we are reliant upon petroglyphs, earthworks and cave art to understand these older human communities, these peat bogs provide us with clues as to the cultures of early human societies. Bog finds help with refining our knowledge and sharing ideas of human agency (Turner and Scaife 1995), informing the interpretative visitor centres now a mainstay of many English wetland spaces (see Wicken and Flag Fen; Somerset Levels' Avalon Marshes Centre as current examples).

Julia Blackburn in her recent work dwelling upon the deep-time antiquity of the East Lincolnshire coast in *Time Song: Searching for Doggerland* (2018) considers the land bridge across the North Sea. Known as Doggerland, the land bridge enabled the flow of humans, and other animals, from across what is now Northern Europe. Blackburn's work alerts us to the ways in which ancient human communities lived alongside, and peripatetically utilised, chains of wetlands across vast land masses. Thinking in terms of 'English' wetlands is to simply make use of a tiny snapshot in geological time. Long ago our contemporary landscapes were shared with now extinct more-than-human brethren too—the mammoths, sabre-toothed tigers, hippos and aurochs whose fossilised skeletons are found across the British Isles (Alleyne 2008). Peat bogs freeze and compress time; they are time capsules, the repository of deep-time memory. Artefacts, organic and abiotic, placed in English wetlands by Mesolithic, Neolithic and Bronze Age humans share similar patterns across the wide continental shelf of Europe (Gearey and Chapman 2004), reminding us that our ancestry is global in scale. Edgeworth (2014) argues that this deep-time vista can be used by archaeologists to communicate climate change science to a wide contemporary audience from an Anthropocene-centric perspective. Wetlands then should be repositioned not as memorial spaces of past human agency but as active learning spaces for the future (see Chap. 6) (Image 5.2).

This alerts us to shared cultures of wetland practices for hundreds of thousands of years across vast terrains. Archaeologists still have a wide range of interpretations about the placing of objects in bogs and the development of monumental structures (Shanks and Tilley 1987; Fowler 2004). Many suggest that wetlands were viewed as the portal to the otherlife, or afterlife, and so placed objects were gifts to the ancestors, working alongside barrow practices such as revering ancestor's bones to mark the changing year at winter, spring, summer and autumn equinoxes (Menotti 2012). Some have assessed the causes of death as ritual sacrifice, with equal weight placed upon theories of deity offerings and of community-sanctioned retributive murder (Chapman and Gearey 2019). These practices span so many tens of thousands of years, with the Stone Age, *lithic*, gauged as a transition lasting 2.5 million years, that we must be guarded in many grand claims, and instead try to reimagine cultural transitions during these epochs. Most of the retrieved items from peat bogs are Bronze or Iron Age, though recent explorative archaeology in the southern North Sea is uncovering items of greater antiquity (Blackburn 2019) including preserved footprints of groups of Neolithic travellers. It is impossible to ascertain how many items have been lost over these vast time periods—with recent retrieval and preservation methods only enabling a very small fraction of these, and bog artefacts, to be saved over the last hundred years.

The plethora of ancient trackways, platforms and marsh forts that have been retrieved highlight the importance of wetlands for ancestral livelihoods and the sustenance gained from fish, wildfowl and plant life within these ecosystems. Platforms and marsh forts found in many wetland systems suggest that nomadic farming practices were operational, enabling

Image 5.2 Doggerland. (Credit: Maxim Peter Griffin)

communities to live on the wetlands and store goods (Norton 2019). A key example of such a trackway is the 'Sweet Track' found by engineer Roy Sweet in 1970 (see Chap. 2) on one of our case study sites, Shapwick Heath (Coles et al. 1973). The small section of perfectly preserved trackway linked with other local archaeological finds, particularly flint axes and antler spears, and links with other, much older, trackways on the site. In particular 'Cheddar Man' found in Gough's Cave in Cheddar Gorge, although only some ten miles from the Sweet Track, is separated more by measurements of eons of time rather than land distance. Cheddar Man's genetic profile indicated that his DNA origins can be sourced from the Anatolian plateau and that human dispersal to the Somerset region could have come through Northwestern Europe. The strong possibility exists then that Doggerland, near our Alkborough case study site, was a means of land based migration. We gain a sense then not only of the integrity of key elements of these 'English' landscapes over deep time, but also the importance of these chains of wetlands for human communities. Alongside sustenance are the commemorative practices of these elder human ancestors.

The work of archaeologist Clive Jonathan Bond in the Somerset Levels alerts us to the adaptivity of humans in response to chages in the Earth's climate. His analysis of 'lithic scatters', evidence of community industry in the Meso and Neolithic, and pollen data, revealing plant life and hence climate and topography profiles, advances a theory that the 'Postglacial Marine Transgression' (Bond 2009: 713) leads to the Sweet Track area morphing from rich woodland to saltmarsh due to seawater inundation during periods of intense flooding across the surrounding drainage basin. As a response, domestic farming moved away from the flatter areas of Shapwick Heath, upland into the adjacent Polden Hills. Rather than completely abandon the site, Bond (2009) suggests that this area became a votive landscape, one in which 'water cults' formed. Possibly in awe of the vast water inundation and land change which occurred, Bond argues that the Sweet Track was only used for ceremony in the later Neolithic, not as infrastructure. He suggests this attends to intergenerational place connectivity—that social constructions of place can lead to working spaces transforming into social, memorialised spaces. Coles and Brunning (2009) proffer a different interpretation, one within which the ancestors were very much incorporated into the fabric of the everyday, such that memory, commemoration and the quotidian might be viewed as collapsed together through practices which have a very different conceptualisation of time and lived lives.

This transformational aspect of lived environments can also be seen when we look at archaic wetland practices in one of our case study sites, Alkborough Flats. Here the dominant feature from antiquity is the 'Roman' turf maze, Julian's Bower (see Image 2.14), which affords the visitor a fine vista of the wetlands beneath from its perch on the limestone escarpment. Turf mazes are found across England, often in sites of Roman occupation (Russell and Russell 1991). It is suggested by historians that Julian's Bower is medieval in construction, but based on the former site of a Roman maze. Further to this, it could be suggested that the maze is pagan in origin, as the Romans often overlaid older sites and ceremonial spaces and practices through imposing Christian reinterpretations of use to eradicate vernacular practice (Heslop et al. 2012). Dinnin and Van de Noort (1999) suggest that Roman occupation of the Humberhead Levels took place when sea levels had receded; so these spaces may have been viewed as new environments, ripe for inscribing the land with the artistic eye of the conqueror. Julian's Bower overlooks Alkborough Flats on an elevated escarpment, providing the space with both views over the Trent and Humber and the feeling of arrival, sited as it now is at the end of the coastal realignment breach. Behind the maze sits Alkborough village church which has a facsimile of the labyrinth in its nave, inscribed into the stone floor (see Image 5.3). This double reflection of the image, in stone

Image 5.3 Alkborough parish church's stone labyrinth. (Credit: Mary Gearey)

and in plant form, provides a biotic/abiotic counterpoint—one which has longevity and the other which requires the constant renewal through the human agency of reshaping and nurturing the growing turf.

Mike Pearson's 2011 performance piece 'Warplands' is focused on his response to Alkborough Flats. Named after the traditional managed floods of farmland, enabling nutrient-rich riverine silt to fertilise agricultural fields and hay-making meadows (known as 'warping' in Lincolnshire and 'drowning' in Suffolk, Norfolk and Sussex as discussed in Chap. 2). Pearson's work is part an oral history, collecting and deploying the use of vernacular terms, mixed with his immediate response to the space. A mix of memory capturing and memory making, he calls attention to this remote, overlooked wetland.

Within the work he draws upon the influence of the river systems in local myth making. The influence of the tides is clear as the estuary, the Humber, meets the freshwater Trent in this space. Pearson refers to the esoteric nature of the water's ebb and flow. He references the Aenir—the tidal bore along the Trent which occurs at the spring and autumn equinoxes. These equinoctial tidal bores have traditionally been associated with a wide range of arcane practices across England. Some interviewees recalled how local legend says that a sacrifice must be made to the river to appease it. Named after a Norse god, the Aenir recalls the earlier occupation of the area by the Vikings. These river banks then have been the site of votive offerings over millennia, lambs were the chosen sacrifice, and these riparian spaces retain an eerie majesty even today. These waters are host to deep-time creatures too. They are the haunt of river and sea lampreys; at three feet long they look less like fish and more like monsters of the deep—the stuff of children's ghost stories. Northeast folklore includes the 'Lambton Worm', felt to be based on the lamprey. These folk stories again link with the Tiddy Mun tales (see Chap. 4) of wetland dwellers needing to be cognisant of the need to appease local water spirits. Memories need not always be bucolic (Image 5.4).

Our final case study troubles deep time and reminds us that landscape memory can be lost without the outliers who often operate without credit or notoriety to preserve an appreciation of our local environments. Our Bedfordshire wetlands are such an example. Although both wetlands are relatively recent manifestations, Bedfordshire remains a largely rural county and therefore one which offers opportunities for reading the landscape for its use over time.

Image 5.4 River lampreys. (Credit: https://commons.wikimedia.org/wiki/File:Lamprey_mouth.jpg)

The two Bedfordshire wetland sites are quite distinct in their memory making and remembrance roles. Both are constructed wetlands—with Millennium Country Park (MCP) as outlined in Chap. 2, a stand-alone wetland, developed from redundant clay pit workings and Priory Country Park (PCP) developed on an existing riverine floodplain. MCP's main water feature is Stewartby Lake, developed in the 1960s from a former clay pit. The whole area around MCP, whose site links Bedford and Milton Keynes at Marston Moretaine, is built on reclaimed brownfield sites that were both former brickworks, particularly around Stewartby, and a number of fully utilised and capped off landfill sites. MCP is then a thoroughly modern, artificial site. Evidence exists of Iron Age, Roman and Medieval use of the surrounding land, including a horse's skull ritually placed during the Romano-British period (Cotswold Archaeology 2012, 18–19). Yet there is scant evidence of any current commemorative use of this site and no evidence of the site as a space that has been revered over intergenerational time. The archaeological report seems to suggest era-specific community use rather than continual celebration of human practice, but suggests a need to incorporate the data into understanding landscape use over time in the whole of the Great Ouse Valley.

PCP, located within the floodplain of the Great Ouse in Bedford town, provides more extensive evidence of human occupation. Priory County Park derives its name from Newnham Priory, developed on the site of an earlier church, around 1100 CE. The original Priory walls frame the Northeastern section of the wetlands. Bedford Castle and mound are located half a mile away on the embankment of the River Great Ouse, and the whole area on which PCP is sited is a large floodplain for the river. The whole area surrounding the river, which includes PCP, is a ritual landscape. At the eastern edge of the wetlands, close to where the power station was located, was a Neolithic henge complex, developed over thousands of years and occupied into the Iron and into the Bronze Age. (Bedfordshire Archives and Records Service 2018). As with other similar sites, evidence for multigenerational occupation suggests a strategically important space, within which the river was a central node, used as a conduit for trade and movement. This is also evidenced by the 'Danish Camp', a Viking settlement, three miles upriver from PCP (Historic England 1991).

However, unlike Somerset and Lincolnshire, these are not peat-accumulating wetlands, and so no organic evidence remains of any ritual practice that may have occurred on these sites. Unlike Somerset where human occupation has been woven into the fabric of interpreting the site, in Bedford this pre-historic henge complex, an earthwork topped with a wooden mounment, has been razed and a Tesco supermarket developed on it, with only a floor plaque in the shop's entrance to commemorate Neolithic place making (see Image 2.4). If we compare the Bedfordshire earthworks with the Sweet Track and Julian's Bower, we see that both the latter examples are visible, identifiable as human intervention within the landscape, easily discerned. Earthworks are harder to read, often needing certain cropping regimes or drought events to make them observable (Daley 2018) as time has eroded their relief profiles. Tim Ingold draws upon the writings of Spanish philosopher José Ortega y Gasset to suggest that 'humans are auto-fabricators' (Ingold 2015: 120). In other words, as humans we are indelibly drawn to leaving our imprints on the planet; and likewise we are drawn to the imprints left by our forebears. For Ingold all landscapes are palimpsests of human mark making (Ingold 1993) when viewed over deep time. Many contemporary people then seek to find ritual space to link with the ancestors. The discovery of the Norfolk seahenge in

1998 bears testimony to this (Wallis 2012)—with thousands of visitors drawn to the site before the timbers were removed for preservation and public display within the Bronze Age centre at Flag Fen in Peterborough (Lynn Museum 2020). Pagans and antiquarians campaigned for the reliquary to be kept in situ, even if this meant the inevitable decay and loss of the artefact.

This speaks to the importance of authenticity in landscapes—that for some human agency in saving items from antiquity also means a diminution of that landscape. It begs questions around what the 'right' thing to do is and how these are subjective, encultured decisions made by 'experts' whose own politics of practice, or praxis, are shaped by contemporary norms. The ways in which we 'read' and value pre-historic artefacts are always a reflection of the contemporary values we place upon encultured spaces (Simpson 2001; Schama 1995). It could be argued that an English wetland renaissance is dependent upon helping current users 'read' the landscape to understand the resonance of these human-induced mark making on landscapes, and also to reflect on their own contemporary lives, as advocated by Edgeworth et al. (2014).

Environmental psychologist James Gibson's (1977) concept of 'affordance' is instructive. Gibson argues that our response to landscape is neither purely cognitive nor physical, but a blending of the two. Our responses are encultured, primal, sensory. We interpret spaces as whole organisms rather than as purely instrumental, rational beings. When we wish to consider memorial practices in wetlands, we see how important this distinction is. Our relationship with landscape moves beyond the rational to incorporate elements of the sacred and the divine. These are often places where humans have sought spiritual solace as well as founding livelihoods, with myths, shared memory and commemorative practices, such as building structures and performing rites, a key element in community building.

The work of Maurice Halbwachs is useful in enabling a consideration that memory is not a personal attribute, but a foundational element of collective community building: it is, essentially, a social enquiry. Although Halbwachs' construction of collective memory is that formed through past lived experience, we might also be able to suggest that through experiential engagement with the world, we, as humans, develop an empathy which enables us to connect with other lives when we encounter a particular stimulus. When we think of wetlands then, these landscapes, and the types of human activity engaged within them, help us to deepen this memory making as more of a social, rather than just individual, endeavour. We can make use of Jan Assmann's (2008) reflection on memory making. As he states:

> As a social construction or narrative, the past conveys a kind of connective structure or diachronic identity to societies, groups and individuals, both in the social and in the temporal dimension.

Assmann argues that memory allows humans to orientate themselves in their lifeworld, placing themselves in the contemporary as well as past time and the future. From this perspective memory is a crucial element of being, and becoming human, and connecting with others, both contemporaneously and across time. Our mark making can be ephemeral—footprints, photographs—or more tangible, planting trees or buying memorial benches for loved ones no longer here (see Image 5.5). Ritual landscapes which evoke human civilisations over deep time can be argued then to be an intrinsic part of our life journey, enabling us to place ourselves in a human-centred cosmology.

When we consider wetlands as ritual spaces, utilised because of their liminal qualities through which human-place connectivity develops around

Image 5.5 Memorial bench, Alkborough Flats, March 2018. (Credit: Mary Gearey)

the concept of physical thresholds and social practices, we can put this into context through an exploration, albeit brief, of contemporary practices on English wetlands which are acts of memory making, commemoration, remembrance and celebration.

Wetland Practices Now

The WetlandLIFE project research has highlighted the numerous ways in which wetland spaces are used as commemorative spaces, places in which people come together for physical and spiritual experiences. Activities which unite the wetland case studies are those which bring groups, or communities of people, together. In particular there is an emphasis on family activities, serving both to engage younger people with nature-based activities and to marketise experience-led social activities of weekend and holiday leisure time. These include guided rambles, night-time bat walks (Image 5.6), star-gazing evenings, pond dipping, bug hunting, litter clearing, invasive plant removal, Nordic walking hikes, bird box making, citizen science species counts, wildlife photography courses, basketry and boat building, amongst so many others. These 'memory experiences' are presented as ways that people can socialise with friends, enjoy activities with their families and develop new skill sets and hobbies. The less managed nature of wetland spaces provides an accessible yet removed green-blue space where people experience a wilder element of the 'outdoors'. In some ways this provides people with the tentative first steps of being in nature, especially for urban residents where access, or time and other resources, to experience wilder places is either limited or non-existent. English wetlands then enable a wide range of othering experiences in often unfamiliar spaces.

The respondents who worked with us on the research also discussed intergenerational activities led by specialist users. For those wetlands with open water resources, these family activities shared across age groups include angling, sailing, kayaking and canoeing, wild swimming and, more recently, otter spotting and paddle boarding. Many interviewees reminisce about passing on interests to their children and grandchildren. This is particularly true of the naturalists who have taught their children to identify species of birds, insects, plants and trees, and those mycologists whose own children now act as county recorders and fungi foragers. The bat group specialists in particular seem to pass on their passion and enthusiasm to family members. With many bat species now protected, there is a

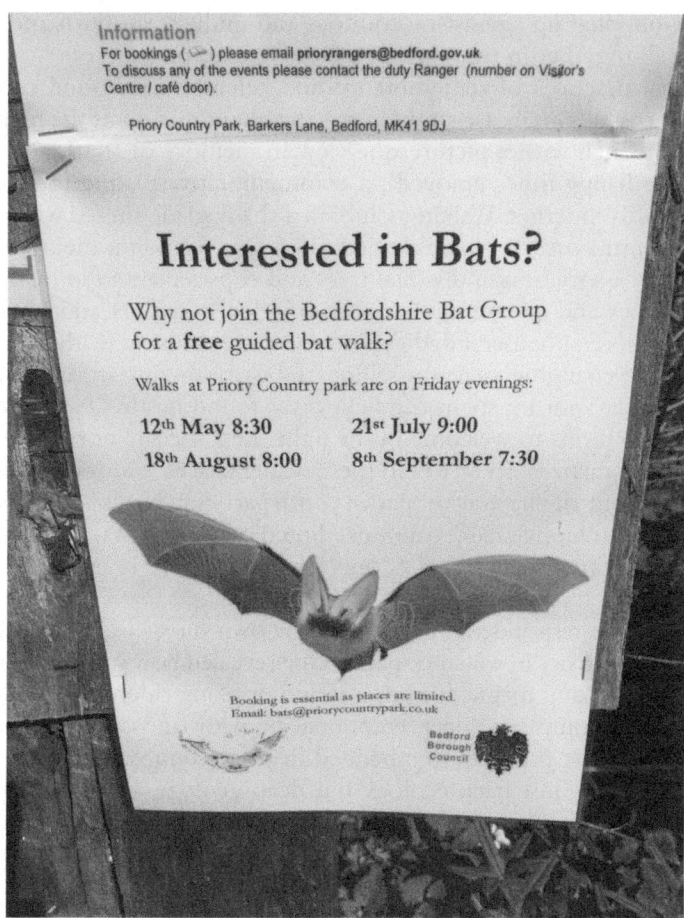

Image 5.6 Family-orientated bat walks. Priory Country Park, May 2018. (Credit: Mary Gearey)

secondary industry developing around undertaking domestic bat surveys as part of planning permission approval.

These remembrance practices are forged then from experiences undertaken in the everyday. Interviewees recalled going with friends and loved ones into wetland sites. This is especially poignant for those interviewees no longer able to access these spaces due to declining health; they still carry with them the fond recollections of hard physical labour, and harsh weather, bundled up against mosquitoes and midges, sunburn and pouring rain, all together in the spirit of collective enterprise.

Material practices of remembrance and celebration abound on these spaces. Across all three case study sites is the phenomenon of the memorial artefact: a bench with a picturesque view in memory of 'Kath and John, and all the happy times enjoyed'; a commemorative plaque to celebrate '25 years of Wanderlust Walking Club' in a shady glade ringed with snowdrops and primroses; a bird hide dedicated to a founding member of an ornithology association; individual trees and copses planted in memoriam to loved ones and as commemorative gifts for weddings and birthdays. There are more ephemeral dedications too: tree dressing in the spring to welcome the changing year; wassailing and carol singing in the winter as the darkest days set in; spent candlelit diyas found nestled bankside after Diwali celebrations to welcome in the light. Less socially normative commemorative practices occur too in these wilder spaces: outdoor raves congregating in out-of-the-way car parks; youth parties in bird hides evidenced with crumpled 'hippy crack' balloons, dinted ends of joints, beer cans and vodka bottles; and discarded condoms hanging off tree branches in memory of wooded trysts.

Some of the respondents talked of their own direct experience with a spiritual connectivity in wetland spaces. One revealed how spending time on the wetlands, taking magic mushrooms and drinking cider with friends has subsequently prompted a deep emotional tie with landscape, so much so that now their art practice is embedded in place connectivity. Their work engages with not just deep ecology but deep geology—seeing landscapes alive in slow-moving geomorphological processes—and are indebted to psychedelic experiences. Michael Pollan in his recent work *How to Change Your Mind: What the New Science of Psychedelics Teaches Us About Consciousness, Dying, Addiction, Depression, and Transcendence* (2019) details scientific evidence which argues that psychedelics promote long-lasting empathetic responses in participants, creating a link with both other humans and the wider natural world, alongside lower rates of anxiety and depression.

Many respondents also attributed improved health and wellbeing through using time spent in the wetlands to reflect upon their own lives. A common refrain across interviews was that wetlands facilitate experiences which close the gap between accessible nature and 'wilder' less accessible spaces. As a result, inhabiting these wetland ecosystems feels like shedding the skin of everyday home and work-life expectations. The changing seasons are most acutely experienced in wetland spaces as there is a tendency for more naturalised planting, rather than the highly sculpted and designed planting schemes found in more urban public park areas, with less tree cover and more exposure to the elements. This enables wetland users to immerse themselves in all the diversity of changing weather and to feel removed from everyday time schedules and commitments. Water is fundamental to the perception of dwelling within these spaces (Strang 2005); respondents talk widely of the importance of getting immersed in the mist in early morning birdwatching forays, of dodging showers running between bird hides, of the inelegant joy of water coming in over the top of walking boots when traversing the moors and heaths. The very particularity of wetlands as big sky, earth and water enables a common feeling of almost being hugged by the landscape, with the enormity of the changing skyscapes above replacing the awesome sublime that mountainscapes afford (Shepherd 2008).

Some interviewees explained how remembering happy times in these places with children when younger, and loved ones now passed, provided solace and respite, replacing potentially nostalgic feelings with ones imbued with love and laughter. Painting, reading, journal writing, thinking and photography are all forms of material/immaterial remembrance practices enjoyed across all the case study sites. One artist referred to the Somerset Levels as almost a solar lamp for the soul, where she soaked up all the feelings, all the sensations of the wetlands to recharge her as she returned to her studio to paint. For those recently retired, the wetlands provide a post-work social habitus—new hobbies and new friendships to ease the transition from one life path to another; commemoration is an active choice, asserting one's own agency in this new phase of life. There is a particular freedom that the wetlands generate—to wander overgrown pathways, to tuck within corners of bird hides as the sky darkens, to close eyes and listen to the sway of the reeds and buzzing of insects. It's almost a churchlike form of reverence—people congregating in these waterscapes to reflect both together and apart, respecting each other's space, yet enjoying communing together as they pass by. Celebration becomes a solo inner journey, as well as a shared one over tea and cake in nearby cafés or on outdoor benches from flasks and Tupperware. The fieldwork enabled one such excursion with a birder, who would choose the same bench that she used to share with her deceased husband, where they would have their post-birding snack. We shared a flapjack and coffee together and talked a little about her husband, and more about the birds she had spotted that year. The birding and the ritual of the shared bench provided solace, an active remembering, in the spaces they had loved as a couple. She brought enough to share, as she always had.

Not all reflections are happy ones. Across all the sites are examples of difficult lives and clashing cultures. Wetland site managers and specialist users commonly voice frustrations at fly tipping, dog mess, dog walking businesses that intimidate other users, illegal campfires and party detritus, fish, game and bird poaching, unlicensed drone flying, industrial scale foraging for restaurants, and other commercial practices. Sympathy is shown for rough campers and the homeless—equanimity towards the young with nowhere to congregate at night. Some voice frustration with visitors who want to park for free and use nearby cafés and picnic benches—but who act roughly towards the ducks and geese who have become tamed by feeding and come too close.

For some the particularly modern curation of wetlands is the source of disharmony. Those who have excavated peat on the Somerset Levels for generations as family businesses are cast as social pariahs as the full extent of carbon losses and habitat destruction has been recognised, and as this industry has been deactivated through non-renewal of extraction licences. Farmers whose lands abut or are downstream from Shapwick Heath and Westhay Moor remember how the landscape used to be, and how the rivers and ditches flowed very differently, before their re-creation as wildlife reserves. They attribute flooding events to these changes as the water levels are viewed by them as being kept artificially high to encourage bird visitations. Many local farmers view these elevated water levels as the cause of localised flooding and a danger for people.

On Alkborough Flats some visitors were unsure of the long-term management of the site, as some viewed the reed beds as too extensive, blocking views of the confluence of the rivers Trent and Ouse, and creating a monoculture that will, they fear, without structured intervention reduce the number of bird species visiting the site. Others suggest that the economic benefits to Alkborough village that were promised by the funded intervention, including a large visitor centre with a café and improved public transport services, have failed to materialise. Some interviewees argued that the wetland focus overwrites other historical uses of the site—such as its role in the Second World War for airborne artillery practice and bombing runs and its use by the Gainsborough pilgrims (see Chap. 2).

In Bedfordshire the legacy of the importance of the power station which helped Bedford to flourish as an expanding market town in the post-war period has been eviscerated in favour of the creation of PCP as a nature reserve. Likewise, it could be argued that the importance of

Stewartby brick works as both an economic powerhouse for the region and as a model village (Stewartby and Kempston Hardwick Parish Council 2020) which provided modestly waged workers with housing, sports facilities and secure jobs, had also been overwritten through the creation of the MCP and the wider Forest of Marston Vale project.

These processes of reshaping landscapes to eviscerate types of heritage which no longer fit current orthodoxy have been described by DeSilvey and Edensor (2013) as deliberate 'ruination'. They ruminate on the ephemerality of human memory: 'Many scholars have mused on the process of ruination as a metaphor for the erosive, unpredictable aspects of human memory' (2013: 471), to suggest that heritage practices are highly selective, highly political processes to present the 'right sort' of local story. Jane Bennett's work on re-enchantment (2001) is useful here—what better way to disguise the ravaged and gouged landscapes of our wetland case study sites than to backfill with lakes, trees, meadows and swans. Re-enchantment also suggests a deep longing for a reimagined past. Through creating constructed wetlands, are we willing to overwrite uncomfortable industrial/military collective memories with ones that fit more comfortably with a 'greenwashed' vision of the future? These tensions are explored further in our final chapter.

Echoing the work of Pendlebury et al. (2017), we might understand these processes as the deliberate obscuring of 'uncomfortable heritage'. Ultimately, Pendlebury et al. (2017) argue, the survival of land and buildings depends on finding new economic uses to replace those that no longer yield a sufficient return. This, they argue, inevitably involves decisions about which stories of the past are carried forward and which are obliterated—a form of what Joseph, Kearns and Moon (Kearns et al. 2010; Joseph et al. 2013; Moon et al. 2015) have termed 'strategic forgetting and selective remembrance'. This is all too evident in our wetland case studies and many other 'remade' wetland sites around the UK. In deciding on the stories to retain, De Cesari and Dimova (2018: 2) argue that landowners and policy makers—aided by international organisations such as UNESCO and the World Bank—assume that heritage and culture are resources that can stimulate local socio-economic development and 'cure' a multitude of social ills. These decisions are, of course, not without values and consequences. The act of making decisions about what to remember and what to forget fosters new possibilities for realising economic gains—for some people at the expense of those who are evicted, displaced or, as many farmers on the Somerset Levels argue, deprived of livelihoods. What we experience, therefore, especially at the Bedford sites, is a form of place-marketing and boosterism (Atkinson et al. 2002) designed to foster economic growth—usually of the development land released by the remaking of the wetland:

> The modern city is marketed as a site of cleanliness, leisure and consumption, but one marked by distinctive historic buildings or quarters, notable leisure or sporting facilities, or natural features or landscapes. Thus, to ensure future economic well-being and social harmony, the message is 'accentuate the positive.' ... Particular aspects of local identity, history or economic activity are marginalised and suppressed due to their negative connotations. For cities scarred by derelict industrial quarters, the legacies of declining, traditional heavy industries characterised by manual labour, grimy production and pollution, the message from place marketeers is to 'eliminate the negative' and focus attention elsewhere, away from these spaces. (Atkinson et al. 2002: 27–8)

New wetlands are the ideal means of achieving this sleight of hand: what could be better than 'swapping' former industrial areas for blue-green, accessible landscapes that point to a bright and sustainable future? This shift from landscapes of production to those of consumption (Atkinson et al. 2002: 28) offers an elision from a past that is no longer attractive to a future that has purpose—a new cultural capital for sites and places that might otherwise have lost their value. But, of course, these new wetland spaces offer limited and selective representations of what becomes a rather singular past in which the dark and uncomfortable have either been removed or have been sanitised as part of a new 'tourist gaze'. For Zukin (1993), these processes are emblematic of the emergence of a symbolic economy based on entertainment, tourism and culture. The redefining of community is significant in achieving this, for while representations of former communities associated with the wetlands are part of the many stories repurposed for tourism, there is often little room remaining for the more recent communities that have often developed in such liminal spaces.

As it stands, the heritage sector is dominated by a particular notion of community, one that overlooks the point that representations of reality can have powerful effects on any group under construction. It can lead to misrecognition, discrimination, lowered self-esteem and lack of parity in any engagement with heritage. This discourse shapes reality, both mystifying and naturalising existing power relations. Importantly, it sets up specific branches of society—heritage professionals and the white middle classes—as somehow devoid of community, existing as nothing more than collections of individuals while 'othering' everybody else (Waterton and Smith 2010: 9).

Memorialisation and heritage in wetlands, especially when linked to the repurposing of place, are clearly shaped by the changing nature of community and the priorities of neoliberal capitalism to create value through consumption, boosterism and the leisure economy. Our respondents in all three wetlands, however, make us reflect that whilst visiting places shaped by the tendencies and contradictions of capitalism, they also perceive themselves as gaining highly personal wellbeing benefits from remembrance, especially those people involved in creative acts that leave material objects on the landscape. Acts of memorialisation can allow people to occupy space and feel mental wellbeing benefits when they are not present, knowing there is a trace in the landscape of a personally valued past (Stenner et al. 2012). The perceived wellbeing effects of the many different ways humans immerse themselves in outdoor spaces mean that many health practitioners now prescribe outdoor experiences, including creative activities, as a treatment for psychological ill-health. Despite this the evidence of the mental wellbeing benefits of outdoor exposure is still much debated (Frumkin et al. 2017). The IPBES (2019) global ecosystem assessment identifies six ways that the range of different ecosystem services can impact positively and negatively on human health, and these are through diet, environmental exposure due to degradation, exposure to communicable diseases, hazards, medicinal products and psychological wellbeing. Having summarised a wide range of evidence, the IPBES (2019) global assessment emphasises that research findings are inconclusive about the effects of nature on mental health although it does conclude that there is good evidence that exposure to the outdoors improves children's wellbeing. This conclusion partly reflects the fact that health risks are influenced by multiple factors, and identifying the scale and features of the contribution of nature to reducing health risks is very complex (IPBES 2019).

Nevertheless, our respondents stress how wetlands as places of remembering and contemplation can inspire and support us in ways that are perceived as beneficial to wellbeing. Simultaneously, the heritage features that stimulate our memories also echo and represent the tendencies of capitalism, thus reinforcing the priorities of contemporary neoliberal agendas. This interaction between personal wellbeing and current forms of capitalism has been at the heart of many critiques of contemporary social and economic organisation. As Mark Fisher (2009) argued a decade ago in the book *Capitalist Realism*, the global mental health crisis has to be understood in terms of how the priorities and requirements of late capitalism now affect not just people's work and finances but also their identity, status, relationships, health and wellbeing. As a result memorial practices linked to the past become less meaningful and their value to individuals reduced. As Fisher (2009: 3) argues, 'in the conversion of practice and rituals into merely aesthetic objects the beliefs of previous cultures are objectively ironised, transformed into artefacts'. All three of our wetlands have heritage and landscape elements that adhere to this argument as artefacts and spaces based on the past are created to attract visitor numbers and expenditure. By contrast, people undertaking acts of personal remembrance along with the practices of artists, writers and wetland managers also create spaces of memorialisation in wetlands that appear to defy commodification and add significant meaning to people's lives. In the final chapter, we seek to draw out these complex contrasts to be found in wetlands as spaces of nature, culture and imagination, shaped by late capitalist society, as we consider the contribution wetlands can and will make within transitions towards sustainable futures.

References

Alleyne, R. (2008) Remains of a sabre toothed tiger the size of a horse found off British coast. The Telegraph 19th November: https://www.telegraph.co.uk/news/science/science-news/3484613/Remains-of-a-Sabre-toothed-tiger-the-size-of-a-horse-found-off-British-coast.html#targetText=The%20partial%20leg%20bone%20of,trawler%20in%20the%20North%20Sea.&targetText=The%20fossil%2C%20which%20is%20between,tooth%20called%20a%20scimitar%20cat.

Assmann, J. (2008). Communicative and Cultural Memory. In A. Erll & A. Nünning (Eds.), *Cultural Memory Studies* (pp. 109–118). Berlin: de Gruyter.

Atkinson, D., Cooke, S., & Spooner, D. (2002). Tales from the Riverbank: place-marketing and maritime heritages. *International Journal of Heritage Studies*, 8(1), 25–40.

Bedford Borough Council. (2020). *The New Stone Age in Bedford*. http://bedsarchives.bedford.gov.uk/CommunityArchives/Bedford/TheNewStoneAgeInBedford.aspx. Accessed on 27th August 2019.

Bennett, J. (2001). *The Enchantment of Modern Life: Attachments, Crossings, and Ethics*. Princeton, NJ: Princeton University Press.

Blackburn, J. (2019). *Timesong; Searching for Doggerland*. London: Jonathan Cape.
Bond, C.J. 2009. A Mesolithic social landscape in south-west Britain: The Somerset Levels and Mendip Hills. In S.B. McCartan, R. Shulting, G. Warren and P. Woodman (ed.) Mesolithic.
Chapman, H. P., & Gearey, B. R. (2019). Towards an archaeology of pain? Assessing the evidence from later prehistoric bog bodies. *Oxford Journal of Archaeology, 38*(2), 214–227.
Coles, B., & Brunning, R. (2009). Following the Sweet Track. In G. Cooney et al. (Eds.), *Relics of Old Decency: Archaeological Studies in Later Prehistory: Festschrift for Barry Raftery* (pp. 25–37). Wordwell: Dublin.
Coles, B., & Coles, J. (1986). *Sweet Track to Glastonbury: the Somerset levels in prehistory*. London: Thames and Hudson.
Coles, J., et al. (1973). Prehistoric Roads and Tracks in Somerset, England: 3, the Sweet Track. *Proceedings of the Prehistoric Society, 39*, 256–293.
Cotswold Archaeology. (2012). *Land at Moreteyne Farm Marston Moretaine Central Bedfordshire Archaeological Evaluation and Earthwork Survey*. https://legacy-reports.cotswoldarchaeology.co.uk/content/uploads/2015/03/660044-Moreteyne-Farm-Marston-Moretaine-evaluation-12066-complete.pdf. Accessed on 1st October 2019.
Cowie, T. (2013). The Ballachulish Goddess: a glimpse of Iron Age religion. *Appin Archive, 37*, 35–37.
Daley, J (2018). Drought Reveals Giant, 4,500-Year-Old Irish Henge. Smithsonian Magazine. July 6th: https://www.smithsonianmag.com/smart-news/drought-reveals-giant-4500-year-old-irish-henge-180969650/
De Cesari, C., & Dimova, R. (2018). Heritage, gentrification, participation: remaking urban landscapes in the name of culture and historic preservation. *International Journal of Heritage Studies, 25*(9), 863–869.
DeSilvey, C., & Edensor, T. (2013). Reckoning with ruins. *Progress in Human Geography, 37*(4), 465–485.
Dinnin, M., & Van de Noort, J. (1999). R Wetland habitats, their resource potential and exploitation. In B. Coles, J. Coles, & M. Schou-Jorgensen (Eds.), *Bog Bodies, Sacred Sites and Wetland Archaeology* (pp. 69–780). Wetland Archaeology Research Project Exeter.
Edgeworth, M. (2014). Introduction. In Edgeworth, M. et al (eds.) Archaeology of the Anthropocene. Journal of contemporary archaeology 1(1): 73–77.
Fisher, M. (2009). *Capitalist realism: Is there no alternative? Zero books*. London.
Fowler, C. (2004). *The archaeology of personhood: An anthropological approach*. London: Routledge.
Fredengren, C. (2016). Unexpected Encounters with Deep Time Enchantment. Bog Bodies, Crannogs and 'Otherworldly' sites. The materializing powers of disjunctures in time. *World Archaeology, 48*(4), 482–499.

Frumkin, H., Bratman, G. N., Breslow, S. J., Cochran, B., Kahn Jr, P. H., Lawler, J. J., Levin, P. S., Tandon, P. S., Varanasi, U., Wolf, K. L., & Wood, S. A. (2017). Nature contact and human health: A research agenda. *Environmental Health Perspectives, 125*(7), 075001.

Gearey, B. R., & Chapman, H. P. (2004). Towards realising the full archaeoenvironmental potential of raised (ombrotrophic) mires in the British Isles. *Oxford Journal of Archaeology, 23*(2), 199–208.

Giblett, R. (1996). Postmodern Wetlands: Culture. In *History, Ecology*. Edinburgh: Edinburgh University Press.

Gibson, J. J. (1977). The theory of affordances. In R. Shaw & J. Bransford (Eds.), *Perceiving, acting and knowing* (pp. 67–82). Hillsdale, NJ: Lawrence Erlbaum.

Halbwachs, M. (1992). *On collective memory*. Chicago: University of Chicago Press.

Heslop, T. A., Mellings, E., & Thofner, M. (2012). *Art, Faith and Place in East Anglia: From Prehistory to the Present*. Woodbridge: Boydell and Brewer.

Historic England. (1991). 'The docks', moated site and dock, Willington, Bedfordshire. https://historicengland.org.uk/listing/the-list/list-entry/1012079. Accessed on 1st October 2019.

Ingold, T. (1993). The Temporality of the Landscape. *World Archaeology, 25*(2), 152–174.

Ingold, T. (2015). *The life of lines*. London: Routledge.

IPBES. (2019). *Global assessment report on biodiversity and ecosystem services of the Intergovernmental Science-Policy Platform on Biodiversity and Ecosystem Services*. Bonn, Germany: IPBES Secretariat.

Joseph, A., Kearns, R. and Moon. G. (2013) Re-imagining psychiatric asylum spaces through residential redevelopment: strategic forgetting and selective remembrance. Housing Studies 28 (1): 135–153.

Kearns, R., Joseph, A. E., & Moon, G. (2010). Memorialisation and remembrance: on strategic forgetting and the metamorphosis of psychiatric asylums into sites for tertiary educational provision. *Social & Cultural Geography, 11*(8), 731–749.

Lynn Museum, (2020). *Seahenge*. https://www.museums.norfolk.gov.uk/lynn-museum/whats-here/seahenge. Accessed on 1st October 2019.

Menotti, F. (2012). *Wetland archaeology and beyond: theory and practice*. Oxford: Oxford University Press.

Mitsch, W. J., et al. (2009). *Wetland ecosystems*. London: John Wiley & Sons.

Naess, A. (2008). *The Ecology of Wisdom*. Great Britain: Penguin.

Norton, S. M. (2019) Assessing Iron Age marsh-forts (Doctoral dissertation, University of Birmingham).

Pearson, M. (2012). Warplands: Alkborough. *Performance Research, 17*(2), 87–95.

Pendlebury, J., Wang, Y.-W., & Law, A. (2017). Re-using 'uncomfortable heritage': the case of the 1933 building, Shanghai. *International Journal of Heritage Studies, 24*(3), 211–229.

Pollan, M. (2019). *How to change your mind: What the new science of psychedelics teaches us about consciousness, dying, addiction, depression, and transcendence.* London: Penguin Books.

Russell, W. M. S., & Russell, C. (1991). English turf mazes, Troy, and the labyrinth. *Folklore, 102*(1), 77–88.

Schama, S. (1995). *Landscape and memory.* New York: Routledge.

Shanks, M., & Tilley, C. (1987). *Re-constructing archaeology: theory and practice.* London: Routledge.

Shepherd, N. (2008). *The living mountain: a celebration of the Cairngorm mountains of Scotland.* London: Canongate Books.

Simpson, M. G. (2001) Making Representations: Museums in the Post-Colonial Era, Second Edition, London: Routledge.

Stenner, P., Church, A., & Bhatti, M. (2012). Human—Landscape relations and the occupation of space: Experiencing and expressing domestic gardens. *Environment and Planning A, 44*(7), 1712–1727.

Stewartby and Kempston Hardwick Parish Council. (2020). *Stewartby History Timeline.* https://stewartbykhparishcouncil.gov.uk/stewartby-history-timeline/. Accessed on 1st October 2020.

Strang, V. (2005). Common Senses. Water, Sensory Experience and the Generation of Meaning. *Journal of Material Culture, 1,* 92–120.

Tilley, C. (2010). *Interpreting landscapes: geologies, topographies, identities.* Walnut Creek: California.

Tuan, Y. (1977). *Space and place: the perspective of experience.* Minneapolis: University of Minnesota Press.

Turner, R. C., & Scaife, R. G. (1995). *Bog bodies: new discoveries and new perspectives.* London: British Museum Press.

Van de Noort, R., & O'Sullivan, A. (2006). *Rethinking wetland archaeology.* Duckworth, London.

Wallis, Robert J. (2012) Pagans in Place, from Stonehenge to Seahenge: 'Sacred' Archaeological Monuments and Artefacts in Britain. In Heslop et al. (eds.) *Art, Faith and Place in East Anglia: From Prehistory to the Present.* Woodbridge: Boydell and Brewer, pp 273–86.

Waterton, E., & Smith, L. (2010). The recognition and misrecognition of community heritage. *International Journal of Heritage Studies, 16*(1–2), 4–15.

Zukin, S. (1993). *Landscapes of power: from Detroit to Disney World.* University of California Press.

Websites

http://bedsarchives.bedford.gov.uk/CommunityArchives/Bedford/TheNewStoneAgeInBedford.aspx accessed270819

https://legacy-reports.cotswoldarchaeology.co.uk/content/uploads/2015/03/660044-Moreteyne-Farm-Marston-Moretaine-evaluation-12066-complete.pdf
https://www.visitcambridgeshirefens.org/wisbech-information-centre-253
https://danishcamp.co.uk/
https://www.explorenorfolkuk.co.uk/seahenge.html
https://www.heritagegateway.org.uk/Gateway/Results_Application.aspx?resourceID=1014
https://ioahc.net/open-day-neolithic-trackway-platform-reconstruction/
https://thepallasboyvessel.wordpress.com
https://stewartbykhparishcouncil.gov.uk/stewartby-history-timeline/

CHAPTER 6

Human-Nature Connectivity: Wetlands Within Sustainable Futures

Abstract This final chapter interrogates what **sustainability transitions** might mean for human-wetland relationships. Utilising contemporary scholarship around **post-humanism**, the chapter explores the ways in which climate change is shaping the way we view and use wetlands. Essential to these arguments are perspectives which draw upon the 'intra-activity' of planetary phenomena through intractable socio-technical 'entanglements'. As our collective future depends on connected adaptations, wetlands are repositioned as essential to sustainable futures. The chapter reflects upon potential barriers and opportunities for wetland expansion, including **biosecurity** issues, **wetland grabbing** and **green gentrification** issues, **nature based solutions** and **green infrastructure** which 'makes space for water' and a cultural reappraisal of landscapes which animate the 'terrain vague'. The book concludes with an overview of the themes of human-nature connectivity within wetlands, amplifying the need to protect these most essential of ecosystems.

Keywords Sustainability transitions • Post-humanism • Entanglements • Biosecurity • Wetland grabbing • Green gentrification • Nature based solutions • Green infrastructure • Terrain vague

Introduction

Standing in an observation hide with eyes closed, in the week leading up to Christmas 2018, all you could hear was the sound of rippling water, like a stream trickling over pebbles. Eyes opened, the vastness of the Pearl River's mud estuary faded away into the heat haze surrounding Shenzhen, a city of over 10 million inhabitants, three miles across the border from the Hong Kong Special Administrative Region (SAR) in mainland China (see Image 6.1), itself home to 7 million people. These wetlands are situated in one of the most densely populated regions on Earth and are highly threatened by climate change and human activity.

Looking out over this section of the Mai Po Wetlands in the New Territories of Hong Kong (Image 6.1), the eye becomes accustomed to a sea of mud, slick and brown in the humid air. The sound of running water came from the mudskippers (Image 6.2), amphibious fish that can live out of the water for periods of time, scooting over the surface of these intertidal wetlands. The mudskippers' ability to change colour to camouflage with the changing light and qualities of the mud surface makes them very hard to see with naked eyes, so the observer hears rather than sees them. They are both a sensorial illusion and an enigma of the animal world, adapted to live in different physical biospheres and across different global

Image 6.1 Mai Po Wetlands, Hong Kong SAR, overlooking Shenzhen, China. (Credit: WWF Mai Po Wetlands)

Image 6.2 Mudskipper. (Credit: sfzoo.org)

regions from the temperate to the tropical. Their adaptive, resourceful, playful characteristics seem so pertinent when we think around the attributes that both wetlands and humans will need to develop as we embrace a planet that is now entering an emergent geological epoch. The Anthropocene (Crutzen 2006) marks the sobering chapter in Earth's history where human-induced activity has radically and irrevocably altered our shared climate; others call for more direct renaming: Capitalocene (Moore 2015), Chthulucene and Plantationocene (Haraway 2015), amongst others.

The WetlandLIFE team visited the mangroves and mudflats of the Mai Po site, and its sister site the nearby Hong Kong Wetlands, in December 2018 as part of a shared learning experience with our Chinese wetland practitioner colleagues, to exchange experiential, scientific and anecdotal learning about our individual wetland research. Objectives included encouraging mutuality in our knowledge development and cultural exchange, and building ongoing relationships of professional support. These wetlands are part of a chain of riverine wetlands, separated by the political border between China and Hong Kong. Though that border is now more porous, the chain link fence and barbed wire which separated (nominally) Great Britain and China still exists; the patrol tower between the two governance regions is still operational (Image 6.3). To visit Mai Po as a non-Chinese resident or citizen, you must sign an authorisation form, stamped by the Chinese border guards. This monumental structure is as potent as the Neolithic trackways, marsh forts, pill boxes and gun emplacements that we have seen in our case study wetlands earlier in the book. These structures serve to remind us of the power regimes in every epoch of human history

Image 6.3 Border control within the Mai Po Wetlands, separating Hong Kong SAR from China. (Credit: Mary Gearey)

that influence how we regard our wetland spaces and the determinants around how we use and access them. Migratory birds, invasive plant species, weather events and rainfall regimes also serve to remind us that sustainable futures reside within the purview of national policy making, but supersede borders too. Sustainable futures are, ultimately, a collaborative venture, especially when addressing climate change.

This final chapter interrogates what sustainability transitions might mean for human-wetland relationships. Key to this analysis is an approach which has at its foundations an appreciation that there is no nature-human divide. From a post-humanist perspective (Barad 2003; Wolfe 2010), our collective future depends on connected adaptations; we are all connected to the 'intra-activity' (Barad 2003: 187) of planetary phenomena through intractable socio-technical 'entanglements' (Swyngedouw 2010). 'Us' includes all humans and our more-than-human brethren (Whatmore 2006). Post-humanism inculcates the perspective that our dominion over nature was only ever ephemeral; in terms of a Lovelockean reading, the Earth, Gaia, will always respond to human intervention as a self-regulating mechanism. Lenton and Latour (2018) suggest that the Gaia principle fails to account for both the extent to which human agency has tipped the Earth past its self-regulating tipping point and the ways in which only human intercedence can help restore a different form of balance. They suggest that humans can utilise contemporary technology to develop systems and techniques to enable flourishing of humans and the more-than-human—a Gaia 2.0. Humans, more-than-humans and artificial intelligence will need to work together to support a collective flourishing on planet Earth.

As we move through 2020, a new decade has welcomed revivified climate justice activism. Greta Thunberg's initiative 'Youth Strike for Climate' has become a global phenomenon. Short-listed for the 2019 Nobel Peace Prize, and author of *No One Is Too Small to Make a Difference*, Thunberg advocates radical change rather than managed transition in response to overwhelming climate change science which suggests humans have superseded threshold tipping points. Tim Lenton, a Professor of Climate Change and Earth System Science, worked on a sister project to WetlandLIFE as part of the UK Research Councils' funded 'Valuing Nature' programme (Valuing Nature 2019). This project *Identifying UK Ecosystem Tipping Points* has had initial findings published in *Nature* (Lenton et al. 2019). Their findings make sombre reading; consequently the team agree with Thunberg's non-negotiable stance. Lenton et al.'s climate modelling accords with the most recent IPCC (Allen et al. 2019) special report on the ocean and cryosphere which has highlighted that low-lying areas across the globe will be subject to flooding due to climate change. Given the global degradation of wetlands, and their critical role in flood storage, we see how vital it is to champion these amazing ecosystems in parallel with behaviour change with regard to fossil fuel emissions. Urgent action is needed: now, not at some agreed date in the future.

Wetlands and Sustainability

Sustainability is an inherent component of these considerations around wetlands. How we contextualise wetlands as a resource, both for their innate biophysical properties and their intrinsic value as spaces for supporting livelihoods in the Global South and increasingly as leisure spaces in the Global North, has direct implications for their visibility in policy arenas. Amplifying the critical contribution of these landscapes to human flourishing has never been more important. In support of this, a crucial element of WetlandLIFE's work has been to explore the historical and contemporary perceptions of these spaces to understand how these changes over time have enabled alternative valuations and uses of these blue-green landscapes.

We have seen throughout the book the ways in which expanding temperate wetlands are imperative for supporting biodiversity and providing climate change impact mitigation resources. Yet there are compromises to be struck. Moving away from the 'drain and reclaim' approach to English wetlands, which peaked in the 1960s, restoring wetlands through arable reversion and coastal realignment, and giving over land to forms of environmental stewardship, including wetlands such as wet meadows and floodplains, from the 1990s onwards has been financially made possible due to our membership of the European Union (EU) and adherence to its Common Agricultural Policy (CAP). Sometimes these EU conservation measures have come into conflict with sustainable land management practices such as Natural Flood Management and Coastal Realignment schemes, as we explored in Chap. 2 Pethick (2002) has highlighted how the EU's 'Special Areas of Conservation' designation can be at odds with interventionist schemes which seek to return pre-drained and reclaimed land back to its former functionality. Brexit opens challenges and possibilities for sustainable land management.

UK agriculture has, through the CAP, been one of the major recipients of funds flowing back from the EU. Indeed, the CAP has shaped British farming, encouraging a bifurcated approach in which prime farmland has been subjected to intense, industrially focused agriculture, while the more marginal areas have had imposed upon them a sanitised version of what rural areas 'should' be. This approach will cease as the UK leaves the EU, whatever deal is struck, leaving up to 50% of farms facing bankruptcy. The government has committed to a phased withdrawal from the current regime, with an emphasis on encouraging older farmers to retire in order that new entrepreneurs can enter the industry. Those remaining in farming are to be incentivised by some public funding being made available for those who provide 'public goods'. While the detail remains sketchy, these are not the public goods of neoclassical economics, but a more market-driven concept in which funding will be tied to some version of public utility. This could certainly encourage the creation of more wetlands, especially those close to urban populations, but the extent to which these will be consistent with supporting complex ecosystems and contributing to broader climate regulation goals will depend wholly on how the funding regime is planned and implemented.

Of particular concern will be the structure and future governance of funding, to ensure not only that public goods are planned but that they are also delivered in the longer term. However, little work has yet been done, it seems, on how the new funding regime will impact on the economics of farm businesses. If it is the case that subsidies shift away from the area-based Basic Payment Scheme, those farmers with large borrowings or substantial rents based on inflated farm incomes are not likely to be in a position to undertake additional work in order to gain new public funding—their businesses will not stand it. In the short term (the next 5–10 years) therefore, the shift away from CAP could see a wholesale decline in environmental schemes being part of UK farming. Eventually, the lower farm incomes that are likely to result from the withdrawal of CAP funding will feed through into lower land prices and farm rents (some studies suggest that rents will need to decline by over 75% in more marginal areas), and those who are left in farming will once again be in a position to plan their businesses to include the creation of wetland habitat and other forms of environmental protection. However, as Chris Rodgers argues:

> If a 'payment for ecosystem services' approach is to be successful … careful consideration must be given to the reform of land tenure law in order to secure a stable platform for the long term management of ecosystems and landscapes. And new legal models will be needed for capturing multi-party obligations for the purchase and sale of ecosystem services. Brexit offers many opportunities for developing a new and innovative approach to agri-environmental policy, but the development of a market for the sale and purchase of the important ecosystem services provided by agriculture will present significant challenges. (Rodgers 2019:38)

Biosecurity: Wetlands, Human Health and Invasive Species

As has always been the case, wetlands are then intimately linked with food and water security. There are also issues of biosecurity, brought about by a changing climate. Biosecurity can be envisaged as a diverse set of techniques and practices in order to reduce, control or eradicate biological threats. Examples include animal passports and quarantine processes (Birke et al. 2013), digital temperature readings of passengers at airports (Warren et al. 2010), and the eradication of infected tree species (Potter et al. 2011). Threats can be orientated towards endogenous systems, where the invasive or predator biological species may change the ecosystem irrevocably, or may be orientated towards regional or pandemic biological threats on human life, or both. This includes plant life, such as Himalayan balsam, Australian swamp stonecrop and floating pennywort (Strong and Burnside 2020), and insect life, particularly those such as mosquitoes and ticks that are vectors for diseases that infect humans. Hinchliffe and Bingham (2008: 1528) broadly define the term thus: 'Biosecurity might be crudely defined as making life safe.' When we consider wetlands as important spaces for sustainable futures, we must also consider the impacts these spaces have on lived environments for other life forms and the ways in which climate change may alter the make-up of these ecosystems.

To consider this in more prosaic terms, the WetlandLIFE project utilised mosquitoes as the emblematic species which might act as the avatar for human-orientated biosecurity issues on wetlands in a changing climate. Mosquitoes are an intrinsic indicator species for wetland health and attendant ecosystem functionality. In both their larval and adult stages, they provide food for fish, birds, frogs, bats and larger invertebrates; they are crucial pollinators in wetland ecosystems. They are also vectors for disease which impacts humans. In tropical and subtropical wetlands, this is evident. Australian government advice (Department of the Environment and Energy 2012) provides information and guidelines regarding symptoms of various mosquito-borne diseases that impact humans negatively—such as *Ross River virus, Barmah Forest virus* and *Murray Valley encephalitis virus* (MVEV), all of which are regionally specific. There are pandemic viruses and diseases too which cross wetland systems, and so political borders. Examples include West Nile fever, dengue fever, yellow fever, malaria and *Zika virus*.

Few people would associate malaria with English wetlands. Yet malaria, and its various colloquially known forms such as marsh fever, tertian fever and ague, has been a historical reality, and the English wetland narrative has been intricately entwined with this disease. Although impossible to substantiate, claims have been made that Oliver Cromwell had malaria when he died in the Irish campaign of the mid-1600s CE (McMains 2015). Hutchinson and Lindsay (2006) have explored historic deaths in coastal marshlands during this period in England and argue that the time of deaths, in mid-summer and early autumn, and the symptoms listed as causes of death point towards malaria as the cause, matching the breeding season of mosquitoes. Jess Beck's detailed photo of an Iron Age skull (Image 6.4) shows us the intricate honeycomb effect that pernicious anaemia leaves in bone material. We know that the most likely cause of pernicious anaemia is malaria which weakens the immune system and lowers levels of iron in infected hosts

Image 6.4 Skull damaged by malarial anaemia. (Credit: Jess Beck)

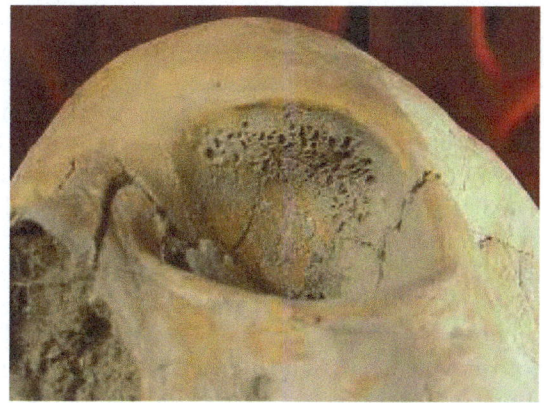

through the destruction of red blood cells. Gowland and Western (2012) use GIS data to map the relationship between Anglo-Saxon skull finds in England with known marsh areas at this time. They explore the hypothesis that skull crenellations, *Cribra Orbitalia*, are manifestations of severe anaemia, caused by malaria, though Schats (2017) argues that as other factors can cause anaemia (such as malnutrition), a more reliable method is to isolate skeletons with blackened bone marrow: evidencing hemozoin, the crystalline byproduct of malarial infection. They conclude that malaria was present at this time in English wetlands.

As a rejoinder to Hutchinson and Lindsay, Watts (2006) argues that it is the artificial construction of water-abundant drains and channels on marshland that increases the density of mosquito populations: human agency has then a part to play in increasing the likelihood of exposure to malaria pathogens as humans manipulate wetland landscapes. Malaria cannot be spread without human-mosquito interaction. We must appreciate the role that humans play in manipulating landscapes which are a contributing factor to biosecurity concerns. Gearey et al. (2000) and Carey et al. (2019) evidence deforestation as a response to emergent settled farming practices in the Mesolithic era as the reason for dramatic and irreversible landscape change. Tree loss changed the soil profile, leading to water retaining areas with high biomass. Over time, and in the geographical space including the area we call 'England', helped along by a warming climate, peatlands formed. This 'deep-time' view of human-landscape dynamics highlights that many contemporary 'natural' wetlands were originally formed through human intervention in the landscape. There is nothing 'natural' about them; but our dominant cultural representations present them to us as such. Iron Age malaria was then a malady directly attributable to human agency.

Sustainable biosecurity contingency planning will need to take into account the non-linear ways in which humans, the more-than-human and a changing climate collide. Whilst there is little contemporary evidence of mosquito-borne malaria infections in England (though there are some that are thought to be contracted via exposure through overseas travel), there is evidence of the possibility that malaria could once again become problematic. Medlock and Vaux (2015) suggest that mosquitoes as vector agents are a distinct possibility, particularly through inter-transmission with infected birds and through ingress through carriage on container ships and other ferry cargo including tourists. This makes English wetlands potentially biohazardous sites (see Image 6.5).

A warming climate also raises the possibility of invasive species and diseases not associated with Northern Europe. The devastation wrought to populations in Brazil by the mosquito-transmitted *Zika virus* highlighted the range and speed of transmission (Centers for Disease Control and Prevention 2016). Swift pesticide responses and other intervention processes are thought to have significantly reduced the incidence of the disease, though evidence gathering from these sites, often rural and dispersed across wide geographical areas, is complex. Evidence has developed to suggest that the changing climate in Southern Europe has led to an increase in mosquito populations—and a greater distribution of species, particularly those known to be the carriers of West Nile fever, yellow fever and dengue fever (Kuhn et al. 2002; Medlock et al. 2012; Fros et al. 2015). There is an anxiety that a lack of public information around mosquitoes and their potential capacity as vectors of infectious disease will lead

Image 6.5 Feeding female mosquito. (Credit: Jolyon Medlock)

to a knee-jerk response to their management. As we have seen, mosquitoes are essential to wetland ecology and are more widely valued for their pollination contribution to other ecosystems. Simply eradicating mosquitoes would certainly have a detrimental impact across the food chain and for biodiversity. Contingency planning seems to be the most pragmatic solution. The WetlandLIFE project has worked closely with Public Health England to both gather evidence regarding extant mosquito species, population and distribution and to understand in more detail how to support and educate different publics about biosecurity issues. In many ways this approach supports a 'rights to nature' (Nash 1989) purview. Eradicating species because they are seen as a threat seems indecent at a time when human-induced climate change is already decimating biodiversity. Instead a more sustainable option is to encourage changes to our social and cultural expectations around how we live with the natural world, thereby opening up our imaginations to creatively explore wetlands as shared spaces.

Flooding, Climate Change, Sustainable Housing and Social Justice Issues

Affirming wetlands as exemplary sustainability landscapes is only part of the complex shift towards forging sustainable futures and adapting to climate change. As argued, part of this narrative is the enculturation of landscape: how we think landscapes should function and what these landscapes should look like. There is still a fragmented public understanding of the significant role that wetlands play in providing rural and urban climate change adaptations. In English wetlands it could be argued that this is due to the emphasis on promoting wetlands as nature reserves, for accessing and enjoying wildlife. This is just one of their many wonderful attributes, but sustainability and adaptation require a full reflection on wetlands within complex socio-natural co-relationships.

Part of the contemporary narrative which involves wetlands at a meta-level is their role in reducing flooding, particularly in urban spaces and as part of wider connected landscapes. River Basin Management Plans (DEFRA 2016a) are used to strategise sustainable use of the wider water environment complex which includes main rivers, smaller tributaries and attendant drainage channels. These plans work alongside UK government Flood Risk Management Plans (DEFRA 2016b) aimed at increasing resilience and reducing community vulnerability to flooding and the 'Making

Space for Water' stewardship initiative (DEFRA 2015) which provides subsidies for landowners and tenants to encourage them to set aside space for water storage. These include a variety of wetland forms including wet meadows, dew ponds, moors and riverine floodplain zones (Evans et al. 2006). These initiatives operate in respect to the European Union's 2000 Water Framework Directive (European Union 2000) and the 2007 Floods Directive (European Union 2007). In the English context, the emphasis is very much on adaptive planning and management to ensure reliable water supply and to reduce peak-time demands which create systemic stresses.

Sustainable water management initiatives which are also reliant upon wetlands are Sustainable (Urban) Drainage Systems (SuDS). Originally developed for low-impact urban housing developments, the benefits of these systems in terms of low cost, low technology and sustainable living mean that they are no longer just the preserve of urban spaces (Butler and Parkinson 1997). SuDS utilises small constructed wetlands to absorb runoff from building exteriors and their internal drainage outputs and use a hybrid mix of plants and soil/aggregates to capture, absorb, filter and purify water entering the system, slowly releasing it back into the environment in multiple ways (see Fig. 6.1). SuDS can be tailored to each individual site, making them adaptable, durable and, in the longer term, affordable. The uptake for these systems is high across Europe, especially Germany, partly because high use drives down price and increases social legitimacy for the product, and partly because of planning and development legislation that requires elements of SuDS in each new build (Dierkes et al. 2015). The uptake in England remains low. Williams and Dair (2007) cite 12 common reasons why SuDS is still not commonly deployed, with lack of implementation skills, sustainability 'trade-off' dilemmas and an inability to enforce amongst the most cited explanations. These findings reinforce that without the political will and financial resources to train and communicate sustainable alternatives, these technologies won't necessarily be part of our sustainable futures. We might, in this context, take a moment to reflect on the estimated £4.2 billion allocated by the UK government to Brexit contingency planning (Perraton 2018) between 2016 and 2018.

Wetlands then play a central role in supporting human wellbeing in sustainability discourses around land management and housing. Their climate change adaptation functionality is an intrinsic aspect when we consider these 'provisioning' and 'regulating' elements of the ecosystem services quadrangle. The WetlandLIFE research data indicated that when

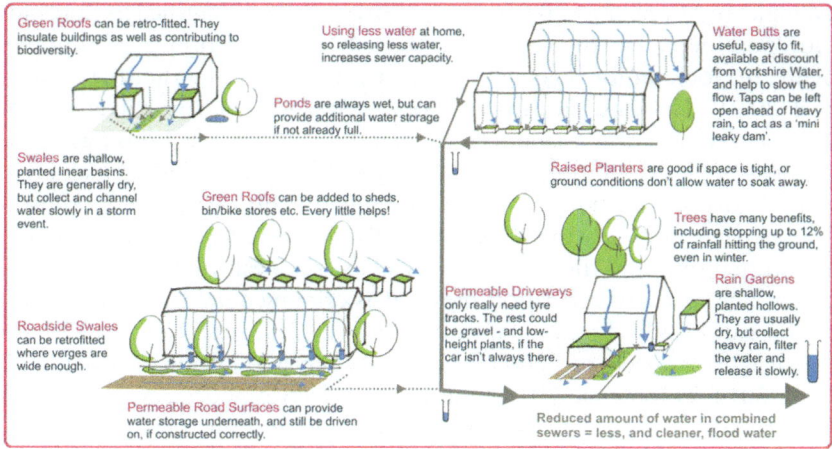

Fig. 6.1 SuDS. (Credit: Slow the Flow Calderdale)

participants were asked to define what a wetland 'is', they invariably depicted less managed spaces—moors, bogs, lakes, meadows. Constructed wetlands were viewed as too artificially managed—too anthropomorphic. Yet increasingly constructed wetlands are utilised to sell urban space. Globally new housing developments are either making use of existing waterfronts or water views or incorporating blue-green space as part of their aesthetic appeal to purchasers.

Human wellbeing is a central promotional tool for these developers (both private and not-for-profit); these green-blue attributes are seen to work in symbiosis with improved human health outcomes and are also a win-win for nature too. Ranganathan and Doshi (2019) call our attention to the ways in which wetlands are used to promote a very particular form of contemporary neoliberal environmental gentrification. They describe a burgeoning global trend of urban wetland development within which property developers build expensive developments alongside floodplains and marshes. This land is bought cheaply, either because it has low value because of its economic unproductivity or in league with regional development interests seeking to regenerate an area. Advertisements attend to residents' abilities to live close to nature. Ranganathan and Doshi (2019) describe these as 'wetland grabs' which gentrify areas, push out low-cost housing and, particularly in developing economies, can create a series of flood management risks. Whilst their evidence base is drawn from examples in Mumbai and Bangalore, this trend can be identified in England too. Woodberry Down Wetlands in central North London is an example of an area looking for regeneration in an era of political austerity (Gearey et al. 2019). Through allowing developers to lead regeneration in favoured waterside areas with guaranteed elements of affordable housing, local councils, with ever shrinking budgets, use these partnerships to support central government housing provision requirements (Monk et al. 2006). These regenerative partnerships are multi-sector and multi-site; but the literature promoting these new homes all use the health-giving benefits of green-blue spaces as recreation sites and as calming aural vistas as positive for human health (London Assembly 2016). Little attention is given to the negative impacts of building on or close to floodplains (Bawden and Clark 2014; Potter et al. 2016), the urban ecology monocultures deployed by such sites (Connop et al. 2016) and the long-term social impacts of creating green-blue space hubs for selected residents only (McGuirk and Dowling 2009). Almost inevitably the affordable and public housing element, in England in particular, is argued to be the bare minimum of provision, reflecting wider policy trends to reposition social housing as almost a 'last resort' status—pushing people into private or quasi-private markets typified by property development companies and housing associations (McKee et al. 2017).

These complex social justice concerns around wetlands as catalysts for gentrification spread to rural spaces too. Golfing and 'executive' estates with constructed wetlands are examples of peri-urban and rural green-blue developments, orientated around highly sculpted landscapes and often incorporating elements of restricted control through security gates, guards or softer defensive landscaping (Blandy 2006). Privileged access to nature, through birdsong, the sound of lapping water and green eyelines, is an intrinsic element of these barbican spaces, where full access is only for those with the appropriate property rights.

Similarly, many new wetlands accessible to the public emerge as a form of privately owned public space (POPS) where private owners, such as water companies in the UK, admit the public to current and former operational sites under strictly managed conditions. In London such wetlands often seek to attract the diverse ethnic communities that surround such spaces (Gearey et al. 2019). As noted in Chap. 3, the ludic activities and practices that emerge in such spaces have socially progressive potentialities that need to be recognised. Sumner and Mair (2017) have noted how 'sustainable' leisure activities can provoke the emergence of new forms of civic commons that brings meanings to people's lives and over time will result in more inclusive public spaces.

The privatised nature of many constructed wetlands, including the new POPS, can distract us somewhat from the bigger sustainability picture. Focus is placed on individual wetlands, rather than connecting them with landscape mosaics and the impacts of climate change across the country. Partly this is to do with the fragmented nature of wetland ownership—with some belonging to wetland trusts, others to councils and many to smaller charitable trusts. Although 'Wetland Vision' (2019) goes some way to remedying this lack of connectivity, the dearth of a national wetland champion impacts on how wetland spaces are promoted to the English public(s). There is still much work to be done to argue for English wetlands as a public good, landscapes which benefit humans at a grander scale due to the wide range of their ecosystem services contributions at the macro scale, such as improving soil, water and air quality, regulating hydrological regimes and enabling biodiversity corridors across wide terrains.

Degrowth, Living Differently and an Aesthetic Re-evaluation of Wetlands

The lack of a unified national wetland vision accords also with a need to address the embodied, aesthetic and sensorial qualities of wetland spaces. As discussed earlier in the book, our case study work shows how prescriptive many interviewees' perceptions are of a 'proper' wetland, even though their functionality remains the same. Eco-critics such as Kate Soper (2007, 2008) and Donella Meadows amongst others (1972) have long cautioned against over-romanticising nature as a cure-all. Both have suggested that post-humanist theory has a tendency to ascribe nature as a salve to modernity: that by recognising humans as nature, a pathway to sustainability will be found. Instead both Soper and Meadows argue in defence of 'human exceptionalism': that humans are possessed of capabilities, distinctive from other animals, to monitor and alter their behaviours in response to changing circumstances. To this end, they further suggest that the most effective way that humans can support flourishing is to recognise the asymmetric impact that consumerist lifestyles have had on the overuse of planetary natural resources. In essence they argue for forms of living differently which are pivoted around living with less, especially weighted towards the Global North as the arena of overconsumption.

Degrowth scholarship and activism are orientated around this concern—learning to live differently in ways which directly usurp the normative foundations of capitalism as the 'natural' order for human society and forms of political economy. Degrowth imaginaries offer alternative ways of envisioning future societies. Attempting to 'decolonise' (Latouche 2015) entrenched presumptions and visualise other ways of being-in-the-world is central to degrowth sensibilities. Degrowth, or *décroissance*, thinking derives from the work of Nicholas Georgescu-Roegen and Serge Latouche and is concerned with human flourishing that relocates 'prosperity' as a global endeavour delinked from monetary goals (Jackson 2009).

A central argument is that the Earth cannot sustain continual natural resources exploitation; sustainable development is, then, an oxymoron. 'Development as progress' is revealed as an unsustainable myth, which has ruined lives and caused environmental degradation. Instead, from a 'post-development' framing, sustainable futures are concerned with learning to live differently—to place wellbeing as the central axis for fulfilling human potential. So what would a 'degrowthed' wetlands look like?

When we reflect on this chapter so far, the importance of wetland expansion, both in rural and urban spaces, for climate change adaptation is tempered by many challenges. As we've seen, biosecurity issues, potential gentrification, or even gated access, catalysed by wetlands and other social justice issues that develop from emergent ways of using and managing water, means that thinking around wetlands and sustainable futures might entail a 'paradigm shift' in the way we consider our future lives both on and with the planet. As we have reflected upon, climate activists such as Greta Thunberg argue that adaptation is not enough. Instead we must end all burning of fossil fuels and radically change our behaviours and actions. A degrowthed wetlandscape might be envisaged as one which utilises thinking from 'Deep Ecology' (Naess 2011) as discussed in Chap. 4, recognising an intimate connectedness between all living things. Wetlands then are not prized simply for their ecosystems services or their contributions to flood water storage or air quality improvement—but are valued for simply existing. All heterodox life forms are recognised as contributing to the connectedness of existence on the planet. As discussed earlier in the work, Michael Pollan's recent work (2019) explores the emerging use of medically supervised ingestion of psychedelics for treating depression and anxiety. The findings seem to suggest that a common experience shared across different medical trials was that recipients felt a deep connection with the natural environment—with common refrains being the experience of understanding our place with other living forms on Earth. The consciousness shift that this entails, and the documented long-lasting wellbeing induced, leads to a reflection on changes in thinking and feeling about our place on the planet. To accept this is to recognise that ways of 'thinking through landscape' (Berque 2013) are a fundamental component of sustainable futures.

EDGELANDS, DECAY AND THE *TERRAIN VAGUE*: DEVELOPING A NEW LANGUAGE FOR WETLANDS

Rehabilitating the image of wetlands has been a crucial aspect of attempts to protect, restore and promote these landscapes. From malaria-ridden swamps, and foreboding marshes and moors through to the 'edgelands' (Mabey and Sinclair 1973; Farley and Roberts 2012) of neither the urban nor the bucolic rural, in European literature at least, wetlands have been firmly placed in the realm of the 'uncanny'. Not just unwelcoming but something more—unproductive, unwanted, unnecessary. As discussed in earlier chapters, these barren spaces unfit for humans, animals or agriculture have, until the mid-last century, been viewed as unsustainable non-spaces, only fit for the mass engineering endeavours of drainage and dredging to repurpose them as productive landscapes.

Clearly times and attitudes are changing towards landscape and its form and function, and as reflected in our project work, these landscapes are now increasingly valued for their importance in a whole range of ways. There is still further to go, particularly if sustainable futures do entail closing the gap between the still self-evident human-nature ruptures. Posthumanism can be argued to be a process rather than a statement. This has been reflected in contemporary cultural representations, as nature writing and popular science have enabled a new and growing audience to see the fundamental importance of letting some spaces just 'be', in all their natural glory. Jorgensen and Tylecote (2007) discuss the ways in which cycles of use and abandonment mark all landscape over time and that, in many ways, wetlands as 'edgelands' found in river corridors and abandoned gravel pits are examples of 'interstitial wasteland' (ibid: 452)—spaces in which to enable counter-cultural practices. They explain these as acts which have no intrinsic 'value'—den building, blackberry picking, fire making, short cuts for waymaking. They praise them as important spaces in which to recalibrate human-nature relations—to slow down the pace and drama of modern life.

Simon Robinson (2017) in his multi-modal work *Archipelagos of Interstitial Ground: A Filmic Investigation of the Thames Gateway's Edgelands*, which is part text and part ethnographic filmmaking, argues for a way of, if not embracing, at least contemplating the worth of urban blue-green landscapes which are unnerving—which do not easily welcome humans. Robinson's ethnographic immersion into these spaces provides ways of settling in to an unsettling landscape of hard concrete edges, wide waterway and floodplain. His films show foxes and crows grooming themselves under streetlights, and urban anglers fishing on scrubby riverbanks. Sustainable futures will involve engaging with edgeland wetlands as well as those more user-friendly and sculpted by visitor centres and information boards. Scholars working in the field of urban design, architecture and environmental psychology have a term for such 'interstitial' spaces in which functionality is ill-defined, in which ownership or forms of use are muddled—*terrain vague* (Gandy 2013; de Solá-Morales Rubió, and Levesque 1995). Theorists such as Mariani and Barron (2013) and Millington (2015) suggest that these spaces, such as the muddy spaces under road flyovers, canalised culverts and abandoned railway lines, are places in which degraded landscapes can be collaborated with, rather than 'colonised' (Stevens and Adhya 2014). Humans and landscapes share agency as we consider future pathways towards emergent forms of living together. How we do this in practice will be shaped by our imaginations, informed by our willingness to step outside of the current capitalist paradigm which still sees nature as operationalised for human benefits only.

As explored throughout the book, we must attend to considerations concerning landscapes and human agency. Wetlands have been used by humans in many different ways over time. Some have always been wetlands, some inadvertently created in the Holocene through the transition from nomadic to settled farming; many have been reclaimed from coastal plains or constructed to replace other landscape forms or use. Certainly within England, which has been the geographical focus of the WetlandLIFE project, expanding urban spaces to create constructed wetlands from former brownfield sites or old industrial estates, or to replace unsustainable rural practices such as peat digging or aggregate removal, has generated curated 'wildernesses' in places of economic decay. Whilst this has enabled the development of wetland sites that attract visitors for the green-blue aesthetic qualities of these spaces, what has been often designed out are references to these former anthropocentric technical practices. In many places it's as if the economic heritage of the site troubles its current form as wetland—and if deemed uncomfortable—is erased completely. Caitlin DeSilvey's work (2006, 2017) has touched on this sensitive issue of legacy and how we wish to represent the present and recent past to our future generations.

This is a fundamental consideration when we engage with sustainable future practices and reflect upon what the legacy of our current 'sustainable development' approaches might be. Recognising the importance of past economic and industrial practices, both negatively and positively, needs to sit somewhat uncomfortably with our modern aspiration to live in accommodation with Earth's systems. Providing visual cues of what went on in these wetland spaces before their reconfiguration is critical in order to remind us of what the Global North is responsible for, what humanity has gained and lost and what more we could lose without more entrenched responses in support of sustainability.

Within the WetlandLIFE project, art has played a pivotal part not just in communicating team ideas and findings but also in helping us as academics and practitioners within the team to reflect on our own relationships with wetlands. Our artists have opened us up to new ways of thinking and doing. Curation of wetlands then doesn't rest just with those responsible for the design, implementation and ongoing management of these spaces, but also with the academic community and the ways in which our own approaches subtly 'curate' perspectives of 'good' and 'bad' practices on these sites. A wider and more heterogeneous engagement with wetlands is needed. Opening out interpretations by enabling visitors to be curious within these spaces, by bringing back in the recent past through physical representations of foregone activities, could be one such method. Using examples from WetlandLIFE's work might mean creating one of the Bedford power station's cooling towers remade with in situ wetland materials such as willow or reed on its former location, lining a drainage bank with antique hods in the Somerset Levels in reference to former peat extraction, creating a sculpture out of used artillery to recall the aircraft training sites in the Second World War at the Alkborough Flats wetlands. Embracing heritage openly within these wetland spaces could, finally, provide them with the voice denied to them for so long.

We end on a positive note. We hope that for some readers this book has enticed you to re-evaluate your own relationships with wetlands, and that when you take time to visit these ecosystems, you may view them just a little differently. We encourage you to take time out to visit these spaces, rural and urban, and to question what is presented to you, to find out more about the human agency behind the sculpting of these sites and the enculturated expectations you bring with you. We also encourage you to slow down, immerse yourself in the wetland elements, make time for your health, your wellbeing in these wonderful spaces. Close your eyes, and let the wetlands, and your imagination, take you on a journey.

References

Allen, M. et al (2019) Technical Summary: Global warming of 1.5°C. An IPCC Special Report on the impacts of global warming of 1.5°C above pre-industrial levels and related global greenhouse gas emission pathways, in the context of strengthening the global response to the threat of climate change, sustainable development, and efforts to eradicate poverty. IPCC.

Barad, K. (2003). Posthumanist performativity: Toward an understanding of how matter comes to matter. *Signs: Journal of Women in Culture and Society, 28,* 801–831.

Bawden, T. & Clark, N. (2014) Why do we insist on building on flood plains? The Independent, 2 February 2018. http://www.independent.co.uk/environment/nature/the-more-the-experts-warn-against-the-more-we-build-on-floodplains-9101710.html. Accessed on 6th June 2019.

Berque, A. (2013). *Thinking through landscape*. London: Routledge.

Birke, L., Holmberg, T., & Thompson, K. (2013). Stories of animal passports: Tracing disease, movements and identities. Humanimalia-a journal of human/animal interface. *studies, 5*(1). https://www.depauw.edu/humanimalia/issue09/birke-holmberg-thompson.html.

Blandy, S. (2006). Gated communities in England: historical perspectives and current developments. *GeoJournal, 66*(1–2), 15–26.

Butler, D., & Parkinson, J. (1997). Towards sustainable urban drainage. *Water Science and Technology, 35*(9), 53–63.

Carey, C. et al. (2020 forthcoming) Identification and analysis of prehistoric brown earth palaeosols under the raised mire of Exmoor. Shared through personal communication 15th September 2019.

Centre for Disease Control and Prevention. (2016) Possible Association Between Zika Virus Infection and Microcephaly — Brazil, 2015. *Weekly* / January 29, 2016 / 65(3); 59–62. (https://www.cdc.gov/mmwr/volumes/65/wr/mm6503e2.htm?mobile=nocontent). Accessed 12th July 2019.

Connop, S., et al. (2016). Renaturing cities using a regionally-focused biodiversity-led multifunctional benefits approach to urban green infrastructure. *Environmental Science & Policy, 62*, 99–111.

Crutzen, P. J. (2006). The 'anthropocene. In T. Krafft (Ed.), *Earth system science in the anthropocene* (pp. 13–18). Berlin, Heidelberg: Springer.

DEFRA. (2015) SW12: Making space for water. Published online 2nd April 2015. (https://www.gov.uk/countryside-stewardship-grants/making-space-for-water-sw12). Accessed 12th July 2019.

DEFRA. (2016a) River basin management plans: 2015. Published 18th February 2016. https://www.gov.uk/government/collections/river-basin-management-plans-2015). Accessed 24th August 2019.

DEFRA. (2016b) Flood risk management plans (FRMPs): 2015 to 2021. Published 17th March. 2016. (https://www.gov.uk/government/collections/flood-risk-management-plans-frmps-2015-to-2021). Accessed 24th August 2019.

Department of the Environment and Energy: Australian Government (2012). Wetlands Australia National Wetlands Update 2012. Issue No. 20, February 2012 ISSN 1446-4843. Advice as at February 8th 2012: https://www.environment.gov.au/water/wetlands/publications/wetlands-australia/national-wetlands-update-february-2012-27.

De Solá-Morales Rubió, I., and Levesque, L. (1995) Urbanité intersticielle. Inter: art actuel (61): 27–28.

Desilvey, C. (2006) Observed decay: Telling stories with mutable things. Journal of material culture *11*(3):318–338.

Desilvey, C. (2017) *Curated decay: Heritage beyond saving*. Minnesota: University of Minnesota Press.

Dierkes, C., Lucke, T., & Helmreich, B. (2015). General technical approvals for decentralised sustainable urban drainage systems (SUDS)—The current situation in Germany. *Sustainability, 7*(3), 3031–3051.

European Union. (2000) Directive 2000/60/EC of the European Parliament and of the Council of 23.10.00 Establishing a Framework for Community Action in the Field of Water Policy–EU Water Framework Directive (OJ L 327, 22.12.2000).

European Union. (2007) Directive 2007/60/EC of the European Parliament and of the Council of 23 October 2007 on the Assessment and the Management of Flood Risks (OJ L 288, 6.11.2007). http://eur-lex.europa.eu/LexUriServ/LexUriServ.do?uri=OJ:L:2007:288:0027:0034:EN:PDF

Evans, E., et al. (2006). Future flood risk management in the UK. *Proceedings of the Institution of Civil Engineers-Water Management, 159*(1), 53–61.

Farley, P., & Roberts, M. S. (2012). *Edgelands: journeys into England's true wilderness*. Random House.

Fros, J.J. et al (2015) West Nile virus: high transmission rate in north-western European mosquitoes indicates its epidemic potential and warrants increased surveillance. *PLoS neglected tropical diseases, 9*(7), 31–36: e0003956. https://doi.org/10.1371/journal.pntd.0003956.

Gandy, M. (2013). Marginalia: aesthetics, ecology, and urban wastelands. *Annals of the Association of American Geographers, 103*(6), 1301–1316.

Gearey, B. R., Charman, D. J., & Kent, M. (2000). Palaeoecological Evidence for the Prehistoric Settlement of Bodmin Moor, Cornwall, Southwest England. Part I: The Status of Woodland and Early Human Impacts. *Journal of Archaeological Science, 27*, 423–438.

Gearey, M., Church, A., & Ravenscroft, N. (2019). From the hydrosocial to the hydrocitizen: Water, place and subjectivity within emergent urban wetlands. *Environment and Planning E: Nature and Space, 2*(2), 409–428.

Gowland, R. L., & Western, A. G. (2012). Morbidity in the marshes: Using spatial epidemiology to investigate skeletal evidence for malaria in Anglo-Saxon England (AD 410–1050). *American Journal of Physical Anthropology, 147*(2), 301–311.

Haraway, D. (2015). Anthropocene, capitalocene, plantationocene, chthulucene: Making kin. *Environmental humanities, 6*(1), 159–165.

Hinchliffe, S., & Bingham, N. (2008). Securing life: the emerging practices of biosecurity. *Environment and Planning, A 40*, 1534–1551.

Hutchinson, R. A., & Lindsay, S. W. (2006). Malaria and deaths in the English marshes. *The Lancet, 367*(9526), 1947–1951.

Jackson, T. (2009). *Prosperity without growth: Economics for a finite planet.* London: Routledge.

Jorgensen, A., & Tylecote, M. (2007). Ambivalent landscapes—wilderness in the urban interstices. *Landscape Research, 32*(4), 443–462.

Kuhn, K. G., Campbell-Lendrum, D. H., & Davies, C. R. (2002). A continental risk map for malaria mosquito (Diptera: Culicidae) vectors in Europe. *Journal of medical entomology, 39*(4), 621–630.

Latouche, S. (2015). *Imaginary, decolonisation of.* London: Routledge.

Lenton, T. M., & Latour, B. (2018). Gaia 2.0. *Science, 361*(6407), 1066–1068.

Lenton, T.M. et al (2019) Climate tipping points; too risky to bet against. Nature commentary 27 November 2019. https://www.nature.com/articles/d41586-019-03595-0.

London Assembly, (2016) Royal docks and Beckton riverside opportunity area. Published online March 2016. (https://barkingriverside.london/, https://www.london.gov.uk/what-we-do/planning/implementing-london-plan/opportunity-areas/opportunity-areas/royal-docks-beckton-riverside-opportunity-area). Accessed 24th August 2019.

Mariani, M., & Barron, P. (2013). *Terrain vague: interstices at the edge of the pale.* London: Routledge.

Mabey, R., & Sinclair, I. (1973). *The unofficial countryside*. London: Collins.
McGuirk, P., & Dowling, R. (2009). Neoliberal privatisation? Remapping the public and private in Sidney's masterplanned residential estates. *Political Geography, 28*(3), 174–185.
McKee, K., Muir, J., & Moore, T. (2017). Housing policy in the UK: The importance of spatial nuance. *Housing Studies, 32*(1), 60–72.
McMains, H. F. (2015). *The Death of Oliver Cromwell*. Kentucky: University Press of Kentucky.
Meadows, D. H., Meadows, D. L., Randers, J., III, W. B. (1972). *Limits to Growth*. Universe, New York.
Medlock, J. M., et al. (2012). A review of the invasive mosquitoes in Europe: ecology, public health risks, and control options. *Vector-borne and zoonotic diseases, 12*(6), 435–447.
Medlock, J.M. and Vaux, A.G. (2015) Seasonal dynamics and habitat specificity of mosquitoes in an English wetland: implications for UK wetland management and restoration. Journal of Vector Ecology 40(1): 90–106.
Millington, N. (2015) From urban scar to 'park in the sky': terrain vague, urban design, and the remaking of New York City's High Line Park. Environment and Planning *A 47*(11): 2324–2338.
Monk, S. et al (2006) Delivering affordable housing through Section 106. Outputs and outcomes. Cambridge: Joseph Rowntree Foundation, University of Cambridge.
Moore, J. (2015). *Capitalism in the Web of Life*. New York: Verso.
Naess, A. (2011). The deep ecological movement: Some philosophical aspects. In R. Bhaskar, P. Naess, & K. G. Høyer (Eds.), *Ecophilosophy in a world of crisis: critical realism and the Nordic contributions* (pp. 96–110). London: Routledge.
Nash, R. F. (1989). *The rights of nature: a history of environmental ethics*. Madison: University of Wisconsin Press.
Perraton, J. (2018) What else could the UK government spend its £4.2 billion Brexit contingency fund on? The Conversation December 20th 2018. http://theconversation.com/what-else-could-the-uk-government-spend-its-4-2-billion-brexit-contingency-fund-on-109110 as accessed on 13th September 2019.
Pethick, J. (2002). Estuarine and tidal wetland restoration in the United Kingdom: policy versus practice. *Restoration Ecology, 10*(3), 431–437.
Pollan, M. (2019). *How to change your mind: What the new science of psychedelics teaches us about consciousness, dying, addiction, depression, and transcendence*. London: Penguin Books.
Potter, C., et al. (2011). Learning from history, predicting the future: the UK Dutch elm disease outbreak in relation to contemporary tree disease threats. *Philosophical Transactions of the Royal Society B: Biological Sciences, 366*(1573), 1966–1974.

Potter, K., Ludwig, C. and Beattie, C. (2016) Responding to a 'flood of criticism': analysing the ebbs and flows of planning and floodplain development in England (1926–2015). Town Planning Review, 87(2): 125–138.

Ranganathan, M and Doshi, S. (2019) Political ecologies of dispossession and anticorruption: A radical politics for the Anthropocene. In: Ernstson, H. and Swyngedouw, E. (eds.) Urban Political Ecology in the Anthropo-obscene: Interruptions and Possibilities. London: Routledge, pp. 91–111.

Robinson, S. (2017). Archipelagos of Interstitial Ground: A Filmic Investigation of the Thames Gateway's Edgelands. How Can a Multimodal (Auto) ethnographic Methodology Be Deployed to Shape Geographic Imaginations of The Thames Gateway? (Doctoral dissertation, University of the Arts London).

Rodgers, C. (2019). Delivering a better natural environment? The Agriculture Bill and future agri-environment policy. *Environmental Law Review, 21*(1), 38–48.

Schats, R. (2017). *Cribra orbitalia: evidence for malaria as the causative agent.* The Netherlands.

Soper, K. (2007). Re-thinking the Good Life: The citizenship dimension of consumer disaffection with consumerism. *Journal of Consumer Culture, 7*(2), 205–229.

Soper, K. (2008). Alternative hedonism, cultural theory and the role of aesthetic revisioning. *Cultural Studies, 22*, 567–587.

Stevens, J., & Adhya, A. (2014). The interstitial challenge: manifestations of Terrain Vague in Detroit and Clichy-sous-Bois, Paris. In P. Barron & M. Mariani (Eds.), *Terrain Vague Interstices At The Edge Of The Pale* (pp. 201–215). New York: Routledge.

Strong, C. and Burnside, N. (2020 forthcoming) The Application of Small-Unmanned Aircraft Systems (sUAS) for the Detection of Invasive Aquatic Plant Species in Wetlands. Shared by personal communication.

Sumner, J., & Mair, H. (2017). Sustainable leisure: building the civil commons. *Leisure/Loisir, 41*(3), 281–295.

Swyngedouw, E. (2010). Impossible sustainability and the post-political condition. In M. Cerreta & V. Monno (Eds.), *Making strategies in spatial planning* (pp. 185–205). Springer, Dordrecht: Springer.

Thunberg, G. (2019). *No one is too small to make a difference.* London: Penguin.

Valuing Nature. (2019). https://valuing-nature.net/.

Warren, A., Bell, M., & Budd, L. (2010). Airports, localities and disease: representations of global travel during the H1N1 pandemic. *Health & place, 16*(4), 727–735.

Watts, S. (2006). Malaria and deaths in the English marshes. *The Lancet, 368*(9542), 1152.

Whatmore, S. (2006). Materialist returns: practising cultural geography in and for a more-than-human world. *Cultural geographies, 13*(4), 600–609.

Williams, K., & Dair, C. (2007). What is stopping sustainable building in England? Barriers experienced by stakeholders in delivering sustainable developments. *Sustainable development, 15*(3), 135–147.

Wolfe, C. (2010). *What is posthumanism?* Minneapolis: University of Minnesota Press.

WEBSITES

https://barkingriverside.london/
https://eur-lex.europa.eu/legal-content/EN/TXT/?uri=CELEX:32000L0060
https://ec.europa.eu/environment/water/flood_risk/
https://valuing-nature.net/

MUDSKIPPER PICTURE – SAN FRANCISCO ZOO. ORG

https://www.london.gov.uk/what-we-do/planning/implementing-london-plan/opportunity-areas/opportunity-areas/royal-docks-beckton-riverside-opportunity-area

https://www.environment.gov.au/water/wetlands/publications/wetlands-australia/national-wetlands-update-february-2012-27. Accessed 040919.

https://www.gov.uk/government/collections/river-basin-management-plans-2015

https://www.gov.uk/countryside-stewardship-grants/making-space-for-water-sw12#features-that-can-be-included-in-this-option

https://www.gov.uk/government/collections/flood-risk-management-plans-frmps-2015-to-2021

https://www.cdc.gov/mmwr/volumes/65/wr/mm6503e2.htm?mobile=nocontent

Index

A
Abiotic, 98, 123, 127
Access, 18, 35, 44, 46, 49, 52, 53, 67, 74, 78, 85, 106, 132, 133, 148, 158, 160
Adaptation, 3, 5, 12, 25, 51, 145, 148, 155, 156, 160
Adolescence, 81, 83
Aesthetic, 9, 17, 18, 44, 54, 57, 70, 111, 140, 157, 159, 163
Affordance, 70, 85, 130
Agro-industries, 57
Alkborough Flats, 22, 33, 47–56, 77, 79, 82, 120, 125–127, 131, 136, 163
Ancestors, 4, 22, 123, 125, 129
Anthropocene, 113, 123, 147
Anthropomorphic, 157
Antiquity, 22, 32, 55, 57, 77, 106, 109, 120, 122, 123, 125, 130
Assemblages, 25, 98

B
Bedfordshire, v, 2, 21, 24, 31–63, 74, 75, 77, 78, 81, 120, 127–129, 136
Biodiversity, 2, 9, 11, 20, 51, 68, 87, 88, 106, 150, 155, 159
Biotic, 2, 3, 98, 127
Blue carbon, 7
Blue-green infrastructure, 48
Bog bodies and bog artefacts, 22, 123
Boosterism, 138, 139
Bronze Age, 123, 129, 130

C
Capitalist, 59, 140, 162
Capitalocene, 147
Catchment, 33, 40, 47, 62
Celebration and commemoration, 55, 119–140
Ceremony, 5, 55, 119–140

© The Author(s) 2020
M. Gearey et al., *English Wetlands*,
https://doi.org/10.1007/978-3-030-41306-4

Chthulucene, 147
Climate change, 2–4, 10, 12, 17, 20, 24–26, 39, 52, 63, 87, 111–113, 123, 146, 148–150, 152, 155–160
Commission of Sewers, 59
Communing, 82, 135
Community Forest, 43, 45, 78
Condoms, 83, 134
Conservation, 9–12, 15, 16, 35, 80, 87, 150
Constructed wetlands, 21, 32, 33, 128, 137, 156–159, 162
Cosmology, 131
Countryside, 13, 44, 46, 83
Crepuscular, 85, 102
Critical theory, vi, 102–110
Cultural capital, 138
Cultural practices, 2, 5, 32, 55, 60
Culture, v, 2, 3, 12, 32, 34, 59, 70, 77, 82, 84, 92, 94, 95, 97, 99, 102, 106, 109, 121–123, 136–138, 140
Curation of wetlands, 136, 163
Curiosity, 5, 70–71

D

Deep time, 1, 4, 17, 22, 24, 83, 106, 120, 122, 123, 125, 127, 129, 131, 153
Degradation, 4, 9–12, 17, 87, 113, 139, 149, 160
Degrowth, 17, 159–161
Degrowth imaginaries, 160
Delinquency, 5
Delinquent, 2, 23, 60, 82, 84, 95, 98
Development, 1, 10, 12–17, 32, 34, 35, 40, 46, 50, 52, 53, 59–61, 73, 76, 87, 123, 137, 138, 147, 151, 156–158, 160, 162, 163
Disorientate, 93, 102

Diwali, 134
Doggerland, 122, 124, 125
Drainage, 12, 24, 41, 56, 58–62, 99, 100, 125, 155, 156, 161, 163
'Drain and reclaim,' 59, 61, 150
Dykes, 60

E

Earthwork, 55, 122, 129
Ecosystem services, vi, 2, 5–6, 9–12, 17, 20, 26, 70, 71, 87, 88, 139, 151, 156, 159
Ecotone, 8, 98, 105
Ecotourism, 58
Eel fishing, 57
Embodied, 63, 75, 121, 159
English culture, 12, 95, 102, 109
Enlightenment thinking, 62
Entanglements, 148
Environmental Impact Assessments, 40
Environmental justice, 3
Epistemologies, 3, 18–21, 32

F

Flood Risk Management, 10, 11, 22, 40, 47, 48, 157
Flood storage, 6, 33, 40, 149
Folklore, 94, 96, 97, 100, 101, 109, 127
Folk stories, 60, 110, 127
Food security, 59
Forest of Marston Vale, 35, 43–46, 137
Fortress conservation, 80, 87

G

Gentrification, 157, 158, 160
Geological time, 122
Glaciation, 4

Grasslands, 8, 24
Gravel beds, 40
Green-blue space, 39, 40, 132, 158
Green carbon, 6–7
Green infrastructure, 40
Greenspace management, 37, 39
Greenwashed, 137
Grey infrastructure, 40

H
Haptic and somatic, 103
Henge, 40, 41, 129
Heritage, 34, 40, 53–55, 96, 121, 137–140
Human agency, 7, 8, 12, 24, 25, 33, 57, 59, 94, 106, 122, 123, 127, 130, 148, 153, 154, 162, 164
Humans and wetlands, 4, 22
Human-wetland relationships, 3, 121, 148
Hybrid, 3, 18, 82, 93, 97, 156
Hydrocitizenship, 18
Hydrosocial, 17
Hyperobjects, 24, 25

I
Illicit practices, 95
Imagination, 2, 3, 8, 21, 22, 34, 70, 74, 86, 96, 98, 102, 105, 109, 140, 155, 162, 164
Indigenous and local knowledges, 10, 11
Instagram, 19, 83, 85
Intergenerational, 2, 81, 125, 128, 132
Internal Drainage Boards (IDBs), 59, 62
IPBES, 9–11, 20, 70, 71, 139

J
Jurassic Ridge, 51, 52

K
Kinship, 18–21, 25

L
Landscape memory, 127
Learning, 22, 70–71, 75, 78, 87, 123, 147, 160
Leisure, 58, 68–78, 80, 121, 132, 138, 139, 149, 158
Levels, 56, 99
Liminal spaces, 72, 84, 98, 120, 138
Lincolnshire Limestone, 51
Livelihoods, 3, 5, 17, 60, 76, 99, 123, 130, 137, 149
Local economies, 3, 86
Ludic, 67–88, 92, 158

M
Magic mushrooms, 95, 134
Managed retreat, 51
Marsh forts, 121, 123, 147
Material/immaterial remembrance practices, 135
Memorial artefact, 134
Memorial practice, 120, 121, 130, 140
Memory, 41, 114, 120, 121, 123, 125, 127, 128, 130–132, 134, 137, 139
Mesolithic, 15, 17, 123, 153
Migrating animals, 7
Migration, 5, 125
Migratory, 2, 24, 33, 148
Millennium Country Park (MCP), 21, 32, 34–46, 75, 78, 128, 137
Mitigation, 2, 3, 16, 25, 150
Modernity, 3, 60, 62, 159

More-than-human, 3, 12, 21, 25, 32, 88, 106, 120, 123, 148, 149, 154
Mosquitoes, v, 134, 152, 154, 155
Murmurations, 21, 68, 85–88

N
National Nature Reserves, 15
Natural flood management, 50, 51
Nature Based Solutions (NBS), 47
Nature writing, 95, 104–110, 161
Neoliberalism, 25
Neolithic, 33, 40, 41, 76, 77, 120, 123, 125, 129, 147
Neolithic raised wooden trackway, 76
North Lincolnshire, 2, 22, 31–63, 78

O
Online, 42, 75, 77, 83, 85–87, 111
Ontologies, 3, 18–21, 32
Oral traditions, 60, 94, 97
Othering, 60, 61, 81, 82, 84, 98, 99, 111, 113, 114, 132, 139

P
Peat extraction, 57, 58, 163
Peatlands, 6–8, 33, 84, 153
Performance, 5, 70, 73, 120, 127
Peripheral economies, 16
Petroglyphs, 103, 122
Phenomenological, 70, 104, 106
Pilgrim, 5, 55, 56, 92, 136
Place making, 129
Plantationocene, 147
Political ecology, 3, 24, 60
Pollinators, 7, 152
Postglacial Marine Transgression, 125
Post-humanist, 148, 159
Post-industrial, 32, 40, 44

Power relations, 68, 69, 99, 138
Pre-modern, 57, 60, 97, 120
Priory Country Park (PCP), 21, 22, 32, 34–46, 69, 73–75, 77, 80, 81, 128, 129, 133, 136
Privately owned public space, 18, 158
Protection, 9–13, 20, 47, 48, 151
Psychedelic, 83, 134, 160
Public transport, 52, 136

R
Ramsar Convention, 3, 10, 16, 17, 21, 25
Raves, 82, 83, 134
Recreation, 10, 23, 44, 67–88, 120, 158
Rehabilitating, 3, 161
Reimagined, 137
Renewable energy, 16
Repurposing, 86, 139
Restoration, 9, 10, 71–73
Restorative, 80
Rewilding, 43
Rhynes, 60
Risk, 40, 46, 51, 139, 157
Romantic, 87, 94, 96
Rough sleeping, 84, 95
Ruination, 137

S
Sacrifice, 22, 84, 123, 127
Safety, 5, 46, 52, 84
Scheduled ancient monument, 55
Second World War, 35, 53, 136, 163
Self-identity, 80
Self-reflective, 94
Sense of place, 70, 81, 103, 107
Sensorial, 146, 159
Sensorial experience, 102

Services, vi, 2, 5, 6, 9–12, 17, 20, 26, 35, 39, 51, 70, 71, 87, 88, 136, 139, 151, 156, 159, 160
Shapwick Heath, 4, 21, 33, 34, 56–63, 75, 76, 78, 79, 124, 125, 136
Social compliance, 100
Social constructions of place, 125
Social diversity, 68
Social justice, 3, 18, 155–160
Social ordering, 61
Socio-environmental transitions, 18
Socio-natural technological change, 63
Solace, 130, 135
Solitude, 79
Somerset, v, 2, 4, 11, 18, 22, 31–63, 76, 78–80, 97, 125, 129
Spiritual, 23, 79, 84, 94, 130, 132, 134
Sublime, 94, 105, 135
Sustainability, 2, 12–18, 148–151, 155, 156, 159, 163
Sustainable development, 16, 17, 160, 163
Sustainable futures, vi, 12, 18, 138, 140, 145–164
Sustainable Urban Drainage Systems (SuDS), 156, 157
Sweet Track, 33, 76, 124, 125, 129

T

Threshold spaces, 23
Tiddy Mun, 100, 101, 127
Trackways and platforms, 121
Tree dressing, 134
Trent Falls, 51
Turf maze, 55, 125, 126

U

Uncanny, 102, 109, 120, 161

Urbanisation, 25
Urban wetlands, 19, 40, 42, 83, 84, 157

V

Vernacular, 56, 61, 99, 126, 127
Visitor centre, 46, 52, 68, 122, 136, 162
Volunteer, 81, 85
Votive, 22, 125, 127

W

Wassailing, 134
Wastelands, 59, 60, 94, 95, 111, 112
Water Framework Directive, 15, 156
Water management, 7, 17, 58, 62, 86, 156
Water resources management, 2, 59, 62
Wellbeing, v, 2, 3, 9, 19, 23, 53, 68, 71, 88, 121, 134, 139, 140, 156, 157, 160, 161, 164
Westhay Moor, 21, 33, 56–63, 78, 136
Wetland grabs, 157
WetlandLIFE project, vi, 22, 32, 78, 84, 96, 132, 152, 155, 162, 163
Wild camping, 83
Wildfowling, 57, 102
Wildlife reserve, 35, 45, 136
Wildness, 42, 43, 120
Wildscapes, 22, 86
Will-o'-the-wisp, 61, 92, 97
Willow coppicing, 57

Y

Youth culture, 82

The manufacturer's authorised representative in the EU is Springer Nature Customer Service Centre GmbH, Europaplatz 3, 69115 Heidelberg, Germany. If you have any concerns regarding our products, please contact ProductSafety@springernature.com

Printed and bound by CPI Group (UK) Ltd, Croydon, CR0 4YY
25/03/2026
02078171-0001